Advanced Sciences and Technologies for Security Applications

The series Advanced Sciences and Technologies for Security Applications comprises interdisciplinary research covering the theory, foundations and domain-specific topics pertaining to security. Publications within the series are peer-reviewed monographs and edited works in the areas of:

- biological and chemical threat recognition and detection (e.g., biosensors, aerosols, forensics)
- crisis and disaster management
- terrorism
- cyber security and secure information systems (e.g., encryption, optical and photonic systems)
- traditional and non-traditional security
- energy, food and resource security
- economic security and securitization (including associated infrastructures)
- transnational crime
- human security and health security
- social, political and psychological aspects of security
- recognition and identification (e.g., optical imaging, biometrics, authentication and verification)
- smart surveillance systems
- applications of theoretical frameworks and methodologies (e.g., grounded theory, complexity, network sciences, modelling and simulation)

Together, the high-quality contributions to this series provide a cross-disciplinary overview of forefront research endeavours aiming to make the world a safer place.

The editors encourage prospective authors to correspond with them in advance of submitting a manuscript. Submission of manuscripts should be made to the Editor-in-Chief or one of the Editors.

More information about this series at http://www.springer.com/series/5540

Franck Guarnieri • Emmanuel Garbolino
Editors

Safety Dynamics

Evaluating Risk in Complex Industrial Systems

Springer

Editors
Franck Guarnieri
MINES ParisTech/PSL
Research University, CRC
Sophia Antipolis Cedex, France

Emmanuel Garbolino
MINES ParisTech/PSL Research
University, CRC
Sophia Antipolis Cedex, France

ISSN 1613-5113 ISSN 2363-9466 (electronic)
Advanced Sciences and Technologies for Security Applications
ISBN 978-3-319-96258-0 ISBN 978-3-319-96259-7 (eBook)
https://doi.org/10.1007/978-3-319-96259-7

Library of Congress Control Number: 2018957612

This Springer imprint is published by the registered company Springer Nature Switzerland AG
The registered company address is: Gewerbestrasse 11, 6330 Cham, Switzerland

General Introduction

The formal study of 'systems' emerged in the nineteenth century, with the birth of industry; around that time, work began to appear of the safety and security of these same systems. Faced with the growing complexity of industrial systems, the modern concept of the 'system' began to be formulated, in various scientific fields, from the second half of the twentieth century.

There are many pioneers, and here we will list only a few:

- Ludwig von Bertalanffy (1901–1972), the Austrian biologist, whose book *General System Theory* has become a reference[1]
- Norbert Wiener (1894–1964), the American mathematician who applied system theory to control and communications[2]
- Claude Elwood Shannon (1916–2001), American mathematician and telecommunications engineer[3]
- Warren Sturgis McCulloch (1898–1969), the American neurophysiologist who broadened his research to mathematics and industrial engineering[4]
- Finally, Jay Wright Forrester (1918–2016), American engineer and professor at Massachusetts Institute of Technology (MIT), who developed the application of system theory to industrial dynamics and who created *system dynamics* at the end of the 1950s, a mathematical modelling technique that makes it possible to understand and analyse the so-called 'complex' problems

Forrester has made a particularly significant contribution. His work and publications have been very well received:

[1] von Bertalanffy L. 1969/1998. *General System Theory*. George Braziller: New York

[2] Wiener, N. (1948). Cybernetics. *Scientific American*, 179(5), 14–19

[3] Shannon, C. (1949). Communication Theory of Secrecy Systems. *Bell System Technical Journal*, 28 (4): 656–715

[4] McCulloch, W. & Pitts, W. (1943). A Logical Calculus of Ideas Immanent in Nervous Activity. *Bulletin of Mathematical Biophysics*. 5 (4): 115–133

- In *Industrial Dynamics*[5] (1961), he describes, with the help of system dynamics, industrial cycles.
- A few years later, he published *Urban Dynamics*[6] (1969) which attracted the attention of urban planners at a global level and led him to meet and join the prestigious Club of Rome.[7]
- From these enriching, productive discussions, he gave us the book *World Dynamics*[8] (1971), which addresses the modelling of complex interactions in the economic, demographic and environmental spheres.

The field of safety studies, like many other domains, could not escape the promise and proven usefulness of system dynamics. In this respect, the work of Jens Rasmussen[9] has had widespread impact. His model makes it possible to study a system by considering its hierarchical structure and dynamic aspects. The integration of dynamics represents a clear turning point in the analysis of accidents and at-risk sociotechnical systems; it allows both negative and positive feedback to be taken into account, thereby creating unique and nonlinear behaviours. Safety becomes a question of 'relations'.

Relations between ideas and concepts (risk, vulnerability, resilience, etc.), between subsystems (prevention, crisis management, feedback from experience, etc.), between man and machine, between organizations (notably in the context of relations between controllers and those they control), etc. It also requires understanding that safety is both organized and organizing. When a company, an institution or a nation produces safety, its constituent elements also act retrospectively on the actions of the entity that created it, by initiating and developing constraints or, on conversely, by creating synergies between subsystems that are constantly changing. Finally, it requires accepting that safety is a potential that actors re-examine and reassess on an ongoing basis, as a function of their needs and hopes, from the point of view of the dynamics and potential of other actors in a given system. This never-ending dynamic can lead to repositioning, evolution, splits and even breaks.

These ideas are generally accepted and therefore widely shared, both within the scientific community and among safety practitioners. However, it is clear that system dynamics has made very few contributions to safety for a very long time. It was not until the work of Professor Nancy Leveson at MIT, a worthy successor to Jay Forrester, that we finally had access to, in the early 2000s, some solid theoretical

[5] Forrester, J. W. (1997). Industrial dynamics. *Journal of the Operational Research Society*, 48(10), 1037–1041

[6] Forrester, J. W. (1970). Urban dynamics. *IMR*; Industrial Management Review (pre-1986), 11(3), 67

[7] The Club of Rome, established in 1968, is a think tank made up of scientists, economists, national and international officials, as well as industrialists, who are concerned about the complex problems facing all societies, both industrialized and developing.

[8] Forrester, J. W. (1971). *World Dynamics*. Wright-Allen Press

[9] Rasmussen, J. (1997). Risk management in a dynamic society: a modelling problem. *Safety Science*, 27(2), 183–213

foundations, a robust methodology and a toolbox for modelling and simulating complex systems[10] (notably thanks to software such as Vensim, AnyLogic, etc.).

Leveson designed and developed a model for risk analysis and accident prevention called STAMP (Systems-Theoretic Accident Model and Processes). The clear, underlying hypothesis of this model is that safety is an emerging property of the system and only exists through the presence of interactions between elements and the application of controls within the system's hierarchical structure (a reference to the work of Rasmussen). Leveson's model therefore represents a radical paradigm shift, as the accident is no longer seen as the result of a chain of events (as is found, e.g. in the Heinrich model[11]) but as the consequence of a control problem within the system.

System dynamics makes many promises. Contrary to popular belief, it is not intended to replace any other forms of modelling nor does try to be more detailed, more precise, more efficient, more effective, etc. Its primary purpose is to invite us to look differently at the world around us. A world that is complex only because we decided it would be. Complexity is not actually a state but an attempt to better describe, understand and share the new knowledge that is acquired through a sustained effort to acquire and formalize data and knowledge in order to produce, present and discuss a result that takes the form of a model.

A model that, through its design process, is in no way a black box but, on the contrary, is an artefact, represented with the help of a diagram, in which it is extremely easy to identify the constituent hypotheses, the descriptive variables and the relations that link them to each other.

The diagrammatic representation greatly facilitates decision-making, in that it offers many new points of view that feed into an evolutionary, iterative and ongoing process. Therefore, even if the model helps to produce imperfect 'decisions', its purpose is to be, at each iteration, better understood and more widely shared. In other words, what is sought is not so much the quality of the choice, as the quality of the process that leads to the agreement to decide. Therefore, the aim is no longer to find the best solution but to be equipped with ways to best manage the uncertainties of the situation in question, examined jointly. To improve the quality of decision processes, the system dynamics approach seeks to clarify and share the viewpoints that led to the modelled situation. It draws upon a dynamic perception of the decision-making process, in which, in particular, the scientific-technical point of view represents only one option, among many others, and which is not assumed to be an accurate perception that the decision must aim towards. The objective is not, therefore, the very ambitious goal of producing decisions and definitive results but that of enriching the decision-making process, whether in technical terms (information, the technical quality of actions undertaken, etc.) or with respect to its sociological aspect (more consultation, giving actors greater power in decisions, etc.).

This book has two aims. The first is to return to the main concepts of system dynamics, put forward a theoretical and methodological framework and describe

[10] Leveson, N. (2011). *Engineering a safer world: Systems thinking applied to safety*. MIT Press

[11] Heinrich H. W. (1931). *Industrial accident prevention: a scientific approach*. McGraw-Hill

rigorous approaches to the formalization of models that are designed to understand and simulate sociotechnical systems. The second is to present some actual industrial case studies, which serve as a basis to illustrate and discuss the applications of theories and methodologies based on data that has been collected from partner companies in the chemical, oil and gas and waste treatment sectors.

This book is structured into two main parts.

The first part is subdivided into two:

- The first, Chap. 1, introduces the concepts of *systems*, the *systemic approach* and *systemic modelling*.
- The second is broken down into three chapters that provide details of the actual implementation of dynamic systems according to the work of Jay Forrester. Chapter 2 shows how STELLA software contributed to the modelling of a chlorine storage facility and its associated risks. Chapter 3 describes the modelling and simulation of human, technical and organizational dimensions in the context of an industrial plant, using Vensim software. Chapter 4 focuses on modelling safety behaviours.

The second part is also subdivided into two:

- The first presents, in Chap. 5, the STAMP accident model and the associated analysis tools: STPA (Systems-Theoretic Process Analysis) for hazard analysis and CAST (Causal Analysis based on STAMP) for accident analysis.
- The second is composed of three chapters that describe operational implementations of STAMP, STPA and CAST: Chap. 6 presents an application to hazardous contaminated sediments; Chap. 7 describes an application to offshore oil installations; and Chap. 8 outlines an application to the hazards associated with the Capture, Transport and Storage of CO_2 (CTSC).

The book ends with a conclusion summarizing the contributions and limitations of the approaches and case studies. Finally, it proposes some avenues for future research.

Contents

1 **The Systemic Approach: Concepts, Method and Tools** 1
Emmanuel Garbolino, Jean-Pierre Chéry, and Franck Guarnieri

2 **Systems Dynamics Applied to the Analysis of Risk at an Industrial Installation** 31
Emmanuel Garbolino, Jean-Pierre Chéry, and Franck Guarnieri

3 **System Dynamics Applied to the Human, Technical and Organizational Factors of Industrial Safety** 93
Hafida Bouloiz and Emmanuel Garbolino

4 **Modelling and Dynamic Analysis of Safety Behaviour** 107
Hafida Bouloiz and Emmanuel Garbolino

5 **Stamp and the Systemic Approach** 123
Karim Hardy and Franck Guarnieri

6 **Using Stamp in the Risk Analysis of a Contaminated Sediment Treatment Process** 151
Karim Hardy and Franck Guarnieri

7 **Contribution of the Stamp Model to Accident Analysis: Offloading Operations on a Floating Production Storage and Offloading (FPSO)** 179
Dahlia Oueidat, Thibaut Eude, and Franck Guarnieri

8 **Systemic Risk Management Approach for CTSC Projects** 197
Jaleh Samadi and Emmanuel Garbolino

General Conclusion 223

Index 229

Chapter 1
The Systemic Approach: Concepts, Method and Tools

Emmanuel Garbolino, Jean-Pierre Chéry, and Franck Guarnieri

The advent of the systemic approach heralded a turning point in the history of science and its application to the organization, and to production. The approach, which considers phenomena and problems as systems, only really began to distinguish itself from the classical analytical approach in the mid-twentieth century, but its origins are much older. The *systemic* approach, as it is currently called, can be considered as a general scientific paradigm, such as the Matter of Life or Society. It offers a generic way to construct and present valid, relevant and rational representations of the most diverse, changing situations. The general system theory, which was conceived by von Bertalanffy (1968), encapsulates these ideas and entails a theoretical and practical method: modelling.

This chapter presents the main principles of the systemic approach. It focuses on the evolution of related concepts, the principal types of models and, more specifically, dynamic modelling. It also presents some examples of the contribution of systems thinking to industry, in particular from the angle of system dynamics.

E. Garbolino (✉) · F. Guarnieri
MINES ParisTech/PSL Research University, CRC, Sophia Antipolis Cedex, France
e-mail: emmanuel.garbolino@mines-paristech.fr; franck.guarnieri@mines-paristech.fr

J.-P. Chéry
AgroParisTech, Montpellier, France
e-mail: jean-pierre.chery@teledetection.fr

F. Guarnieri, E. Garbolino (eds.), *Safety Dynamics*, Advanced Sciences and Technologies for Security Applications,
https://doi.org/10.1007/978-3-319-96259-7_1

1.1 The System and the Systemic Approach

The word *system*, which generally means "assembly", derives from the Greek verb *systeo,* a word that was used to denote the action of attaching elements together. In systems thinking, authors define their own interrelationship between elements, as illustrated in Durand (2006, pp. 7–8) who offers the following examples:

- "A set of units in mutual interaction" (von Bertalanffy);
- "A set of elements in dynamic interaction, organized around a goal" (de Rosnay);
- "A global unit made up of interrelationships between elements, actions or individuals" (Morin).

If the system is considered as an object of knowledge, it can be defined, after Le Moigne (1977), using three characteristics (see Fig. 1.1):

- The *ontological* pole refers to what the object *is*. It considers the system in terms of its elements, and its boundary with its environment. This defines the system's structure, i.e., all of the *a priori* stable relationships between the elements.
- The *functional* pole refers to what the object *does*. This is clearly linked to the nature of the system, and therefore the ontological pole. From the theoretical point of view, the functional aspect is characterised by goals. Beyond consideration of what the system does, it also addresses what the system is *for*. Artificial systems, such as machines or robots,[1] are designed and built by humans. These artificial or technical systems can be easily identified. They have little capacity for change: their goal is clear and defined by their designers. It is more difficult to identify the purpose of open (e.g., living) systems, which raises philosophical and spiritual questions, such as "What is the meaning of life?" In this context, it becomes important to distinguish between the function and purpose of a system,

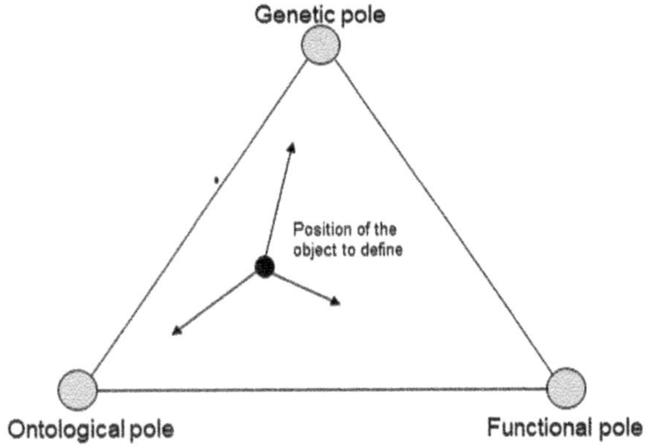

Fig. 1.1 Definition of the object of study using a triangulation based on the poles "Genetics", "Ontological" and "Functional". (Following Le Moigne 1977, p. 39)

[1] The concept of the artificial system extends beyond the boundaries of the finished object: beyond the machine or robot, there is the human designer.

or its *means* and its *ends*. In general, complex systems have multiple ends: potential contradictions arise when trying to satisfy incompatible goals or objectives.

- The *genetic* pole refers to what the object *becomes*. The system is observed over time, and considered to have the capacity to re-define itself over time. From this point of view, memory, history, trajectory and dynamics are system attributes.

The following section looks at the development of systemic thinking in the West, and compares it to analytical thinking.

1.1.1 History

Western scientific thought became autonomous between the late fifteenth century and the twentieth century with the rise of rationalism. It incorporated the scientific heritage of antiquity and medieval thinking in the Mediterranean, through various intermediary countries (Egypt, Greece, Persia, India, and the Arab-Muslim world). In the West, the constitution of a socially-recognized body of scientists, supported by an increasingly specialized body of engineers, emerged in the seventeenth century. It was at this time that Descartes proposed a scientific approach that shattered the scholastic tradition, which was increasingly considered to be obsolete in light of the great Copernican and Galilean discoveries. The words of the Bible could no longer provide an explanation for everything, and it became necessary to construct a method that met the new need for intellectual rigour. In his *Discourse on the Method*, Descartes (1637) outlines four precepts that underpin the analytical approach: doubt and certainty; the division of problems into as many parts as possible; the assembly of simple objects; and completeness.

Analytical thinking postulates that everything is the sum of its parts; likewise, a general explanation is made up of a set of individual explanations. This provided the foundation for one of the key currents of scientific thinking that describes the world. The nomenclature claims to be exhaustive and exclusive: anything can be categorised into only one category, and be placed in a hierarchy of other categories. In the eighteenth and nineteenth centuries, this approach gave rise to taxonomy and systematics in the life sciences, formalized by Carolus Linnaeus (Encyclopaedia Britannica, online Internet site).

However, analytical thinking was not the only school of scientific thought during the long period of Western imperialism. For example, in his posthumous *Thoughts,* Pascal (1670), expressed the holistic thinking that is at the heart of systemic approach.

"Since everything then is cause and effect, dependent and supporting, mediate and immediate, and all is held together by a natural though imperceptible chain, which binds together things most distant and most different, I hold it equally impossible to know the parts without knowing the whole, and to know the whole without knowing the parts in detail."

This quotation indicates the need for an overall vision. Everything has properties that cannot be reduced to the sum of the properties of its parts. This line of thought is characterised by Pascal's firm religious beliefs that support the notion of the divine.

Earlier Greek authors, such as Heraclitus, held atheist beliefs (Patrick and Bywater 2014). For Heraclitus, access to the totality is considered through the union of opposites, their coexistence and co-production: "living in their death and dying in their life". The opposition does not require an explanation based on causal exclusivity—linear causality—and narrow system ontology. Heraclitus also poses the problem of identity, or the ontology of objects, with the famous phrase "Into the same river you could not step twice", which raises the issue of permanence, conservation vis-à-vis dynamics, time and changes in appearance.

It is not until the early twentieth century that the failure of analytical thinking to provide adequate answers becomes apparent. It is faced with a growing body of knowledge that is scattered across highly compartmentalized disciplines in an academic structure that is, on the one hand, inherited from encyclopaedism and, on the other, from the positivism of Comte. This is illustrated by the synopsis of his *The Course in Positive Philosophy* (*Cours de Philosophie Positive*) in 1826 (Le Moigne 1990, pp. 154–155).

The first criticism of the analytical approach or, rather, the first suggestions that another view of the scientific process was possible appeared at the dawn of the twentieth century. Changes were seen during the first third of the century: in physics, with the rise of thermodynamics and its implications for industry; in evolution, in population growth research in biology; and in the use of iterative calculation in differential equations. The issues that were raised in the first half of the twentieth century addressed the notion of different states, potential or realized, and the role of this difference in processes. Wiener and von Bertalanffy (Durand 2006) were important contributors to the founding of a new scientific approach—cybernetics and systems science in the 1940s.

In biology, von Bertalanffy's work does not require the separation of bodies in order to differentiate them, in a dissection-like process. The isolation of organs or the accuracy of microscopic descriptions is not enough to explain what a living organism is. Moreover, the Darwinian revolution of the late nineteenth century launched the search for explanations in terms of relationships between organisms and their environment, their habitat and their population in competition with other types. The emergence of genetics, in the early twentieth century, suggested that living organisms have physico-chemical properties that act as non-material information. In biology, the emergence of new forms of multicellular organisms—from an initial cell, for example—required going beyond the analytical method.

In order to try to understand the contribution of systems theory, de Rosnay (1975) compared the characteristics of the analytical approach and the systemic approach (Table 1.1).

Table 1.1 Comparison of analytical and systemic approaches (After de Rosnay 1975)

Analytical approach	Systemic approach
Isolates	Unifies
Nature of interactions	Effects of interactions
Precise detail	Global perception
One variable at a time	Groups of variables
Reversible phenomena	Duration and irreversibility of phenomena
Validation by experimental proof	Validation through modelling and simulation
Precise and detailed models	Global templates
An efficient way to handle linear and weak interactions	An efficient way to handle nonlinear and strong interactions
Teaching by discipline	Multidisciplinary education
Detailed action programme	Action through objectives
Knowledge of details	Knowledge of goals

As with any beneficial innovation--and, in science with an improved understanding of phenomena—adoption was gradual. The systemic approach was taken up over several decades by other disciplines, countries, research networks and application areas (industry, management, information technology, engineering, etc.).

One particularly interesting contribution of the systemic approach is its contribution to the development of interdisciplinary research. As it connects the components of a system, it can be used to study a system as a whole. It highlights emerging behaviour, and the fact that systems can be seen as a methodological support that links the knowledge, expertise and data from various disciplines relating to the same system. Consequently, industrial risk management has become an interdisciplinary activity: engineers took an interest following the emergence of dependable systems in the 1920s. Since the late 1980s it has integrated approaches and tools from information technology, ergonomics, psychology and sociology—as can be seen in the pages of international journals such as *Safety Science*, *Risk Analysis*, and the *Journal of Loss Prevention in the Process Industries*.

The role of the systemic approach is to define the means of prevention and protection in response to industrial hazards. Despite the lack of methodological developments, there are some examples. The Method Organized for a Systematic Analysis of Risk (MOSAR) procedure was designed in the late 1990s (Perilhon 2003) and is based on modelling the industrial system and a two-level risk analysis. The macroscopic analysis examines the main elements in proximity to each other (operators, hazardous materials, at-risk processes, etc.); and the microscopic analysis explores the technical and operational faults previously identified using traditional dependability methods. The approach proposed in this book is similar to MOSAR, but here the behaviour of the industrial system is studied dynamically, i.e., the evolution of the system's elements and its general behaviour are examined over time.

The following section completes this brief history by presenting the key concepts of systems thinking.

1.1.2 The Main Concepts of the Systemic Approach

Durand (2006) identifies four major concepts that characterise the systemic approach:

- interaction
- comprehensiveness
- organization
- complexity.

These are described below.

Interaction is related to causality in a system: elements interact, i.e., they perform actions on other elements and are subjected to actions by other elements. When one element does not interact with any others, possibly because it only performs or receives one action, it is considered to be external to the system. The systemic approach offers the advantage of diagnosing the causal relationships that describe an element's reflexivity: the action of element *A* on element *B* affects the nature or intensity of the action of *B* on *A*. This feedback occurs in many natural and artificial systems. It can be energetic, material or informational depending on the type of system.

Comprehensiveness reflects the notion that everything cannot be reduced to the sum of its parts. There are specific properties that depend on the subset of the system in question or the entire system. These irreducible properties change depending on the degree of aggregation of elements, or subsets of elements of a system: the hierarchy of the aggregate elements reflects these qualitative or quantitative changes in recognized properties. It is important to include them in the study of a system.

Organization refers to the consideration of both the structure and operation of a system. Typically, the arrangement of the system's components assures its functions and processes. These functions and processes are differentiated in such a way that the elements that make them effective constitute the actual structure of the system. Organization implicitly suggests, in common parlance, a goal. This is generally the case for most artificial and social systems, in the production of goods or services, for example. However, the organization of natural, open systems is a poorly understood aspect that science seeks to explain. For example, zoogamy—the pollination of flowering plants by animals such as insects—is an *organization* of the reproduction of these plants that shows that the *plant system* is open to other living species. The establishment of control within the organization cannot be understood using the simple analogy of a closed, artificial system. To achieve the unalterable goal (if it can be identified) of reproduction, the phenomenon of inter-species symbiosis has to be established.

In economics, the comparison of *intra*-enterprise performance based on an integrated model, and *inter*-enterprise performance based on a decentralized model—such as the *industrial district*—leads to a debate about the modalities of controls in exchanges between the system (the organisation) and its environment, and its ability to achieve its purpose when the system is very open. In risk management, the implementation of a safety management system to monitor subcontractors, for example, is an additional organizational means of control between the company and its socio-economic environment. This safety element, which is typically organizational, is necessary because of the interactions between the company and its contractors; these exchanges can generate failures that can lead to accidents.

Complexity must be distinguished from the term *complicated*. A complicated system consists of many elements, and the multiple relationships make it difficult to understand. The end result, however, may be a simple, stable, repetitive cycle that runs like clockwork. The analytical approach can be used to understand this complicated system. Generally, a complicated system has linear causalities, little interactive causality and is unlikely to be open to its environment. There are several properties or characteristics of complexity, however, that can be observed in some systems (Zwirn 2006):

- Self-organization: the ability of a complex system to change its organization without any causal influence from its environment;
- Emergence: the emergence of new and dynamic system properties, characterized by the concepts of phase transition and percolation.

Other properties (such as the system's sensitivity to conditions or constraints on its subsequent dynamics and its adaptability) demonstrate complexity. Generally, complexity suggests that it is difficult to predict the dynamics or evolution of a system (Donnadieu and Karsky 2002). Complexity can be viewed as "uncertainty in richly organized systems" (Morin 2005, p. 49).

Unpredictability can be reduced by taking into account those elements that were initially excluded from the system, but which are subsequently found to have strong causal relationships with those items that were initially included in the system. The aim is to take better account of a system's spatial complexity—*spatial* in the sense of the structural relationships between elements—by extending the system's boundaries. Another aspect of unpredictability which is very difficult to reduce, relates to the system's temporal dynamics. Certain temporal phenomena (dependent on the system's spatial complexity) produce events that can create bifurcations in the system dynamics.

These four major concepts that underpin the systemic approach reflect the difficulties of studying, understanding and acting on natural and artificial systems. One method that can be used to assess, diagnose and understand these types of system is systemic modelling.

1.2 Systemic Modelling

Modelling a system involves constructing a representation of the system—a model—that simplifies, at least in part, its structural and functional properties. The representation seeks to maintain, as far as possible, intelligibility, reliability and usefulness. Several categories of system models can be identified.

1.2.1 The Major Categories of Models

Research in systems modelling has led to systems being classified in several different ways. One example, based on the level of organization of systems, defines nine levels that constitute what Boulding (1956) calls the *general system*. These levels are represented, in increasing order of complexity, by: (i) the frameworks (the static structure); (ii) clockworks (a simple dynamic system); (iii) the thermostat (the control mechanism, or cybernetic system); (iv) the cell (the open system, or self-maintaining structure); (v) the plant (the genetic system, differentiated functions); (vi) the animal (the decisional system); (vii) the human (an intelligent system that is able to imagine, interpret symbols, etc.); (viii) social organization (socio-cultural system); and (ix) transcendental systems (systems that exceed the capacity of human knowledge to understand, but which humans can question).

Based on his modelling method and a re-reading (and adaptation) of the work of Boulding (1956), Le Moigne suggested (1977, 1983 pp. 128–149) that an organization can be made up of the following new levels:

1. The passive, unnecessary object: these objects are what they are, they are considered in their entirety by the modeller. They include, for example, atoms, planets, stars, cells, words in isolation, an alarm, a valve, an operator, etc.
2. The active object: the object moves and can act on its environment, other objects, etc. For example, the planets move around the sun, the latter's gravity acting on the former. This activity does not change the nature of the object. The activation of an alarm does not change its nature or the elements of the system, but may trigger some of them to react.
3. The active and regulated object: determinism suggests that its behaviour appears to be related to initial conditions. Similarly, its behaviour is regulated by control loops that give it stability. For example, the activation of an alarm triggers a servo valve to close, and in turn, the alarm stops.
4. The object is aware of itself: the object is open to its environment and communicates with it to maintain itself in a stable condition. Here, information becomes an element that links the object with the world around it. The object perceives, and can represent its environment. Leakage sensors positioned around a storage system provide electronic devices with information that can activate or deactivate an alarm, if needed.
5. The object decides what it does: the object demonstrates an internal logic that drives it to make decisions, based on a goal, as a function of its perceptions and environmental constraints. Systems that are serviced and controlled by software can be activated as a function of changes in environmental parameters.
6. The active object has a memory: this memory is closely associated with the decision-making process that integrates the representation and information flows in its environment. Le Moigne uses this to introduce memory processors that are linked to decision-making processors. Computer-based control systems used in industry can record information captured by sensors and interpret it. For example, software can anticipate thermal runaway in a chemical reactor.

7. The active object can coordinate: here, it acquires the ability to coordinate its activities given the information available to it. Some robots are able to work independently in an extreme environment to gather information on the state of a system.
8. The active object can imagine and self-organize: the object has the ability to generate symbolic information that may not be directly related to the information it perceives. It can develop a new logic that the decision processor can use to adopt a new behaviour. This level is characterized by the emergence of intelligence. The operator or group of operators is able to organize and adapt to a new situation—for example crisis management on an industrial site.
9. The active object can autofinalise: the object is able to formulate projects, for itself, its entourage or its environment. This level equates to the emergence of consciousness (self-awareness) in relation to the environment (the object's place in the world around it). Any at-risk industrial project is found at this level as the decision-making process is based on multiple criteria (economic, environmental, ecological, social, etc.).

Le Moigne's proposed control system consists of the finalization system, the intelligent system and decision-making system. This control system interacts with the information system and the operating system.

More recently, Durand (2006) adopted a different perspective, and developed a taxonomy of models based on how they are used:

- The *cognitive model* is a simplified representation of the system, which focuses on the system's knowledge. Typically, cognitive models use graphs and pure analogue forms such as symbols. An ellipse can be used to represent the path of the Earth around the Sun, for example, where the line represents the path of the planet around its star that, in simple terms, cannot be departed from. Similarly, in industry, Piping and Instrumental Diagrams (PIDs) are used to provide a simplified view of the systems implemented in a process; and in dependability, the Bow-Tie risk analysis method provides a simplified graphical representation of the occurrence of a dangerous phenomenon and its impact on the system, the environment and health.
- The *decision model* is a decision-making tool that is typically used for optimization. The decision to be made is expected to be the best possible in a given context. Usually, decision models seek, through simulation, to represent the different consequences of decisions. They then use computer programs based on combinatorial and probability principles. The modelled system exhibits equal-end properties, i.e., given different initial conditions (e.g., initial decisions), the system reaches the same end state. This shifts the choice of the decision from the goal to be achieved, to the cost of reaching that goal (some decisions are more expensive than others). Software packages that can simulate dangerous phenomena play a role in decision making in urban planning and crisis management, for example, through the definition of safety perimeters.
- The *normative model* focuses on establishing parameters to monitor, maintain or achieve, depending on the state of the modelled system. This model focuses on

human and artificial technical systems. A normative model can only focus on the structure of the system or its operation. Normative models may reflect the organization or management of an economic activity, typically a business. Activities that pose risks (threaten the safety of people, property or the environment) in some industries can be identified using models that set safety rules. Threshold limits are established that must not be exceeded, such as the maximum load or concentration of a product in physical and chemical processes; at the same time humans monitor system performance and take appropriate action (e.g., alert procedures and backups). Most accidents occur because the normative model was either poorly designed or poorly applied. In the latter case, it may still be argued that the model was poorly designed, in that it did not take appropriate account of the elements that caused the problem.

- The *predictive model* can produce information that indicates the future state of the modelled system, including in abnormal or unusual conditions. Generally, the predictive model's design uses a dynamic representation of the system in an appropriate past, and present timeframe. From this representation of the structure and the past operation of the system, the (typically computerised) model reports trends in the system state or in scenarios that change the system's dynamics, which provide an understanding of its sensitivity. Given the ever-increasing power of computing technology, predictive models are implemented as increasingly sophisticated computer programs. These models can be deployed in monitoring tools in the chemical industry, for example, where they can be used to anticipate the likelihood of a thermal runaway so that appropriate preventive actions can be taken.

These categories of models are not mutually exclusive. The modelling approach aims to increase understanding, improve reliability and enhance usefulness of the system. It produces a model that reflects these properties. Generally, *a minima*, all models are cognitive: the system that an actor seeks to act upon, to set rules for, or to predict its future, must first provide a representation of itself that expresses the state of its knowledge. In industry, the categories of models that are used depend on the particular system being studied. The codified representation of a given industrial plant or its principle elements (buildings, flows, production tools, control equipment, safety systems, etc.), for example, is simultaneously a cognitive model and a normative model. Similarly, official documentation, such as plans of the industrial facility, typically reflects both of these aspects of modelling.

An industrial plant can also be represented by a decision model that incorporates knowledge of risk factors and accidents, along with the *a priori* actions that need to be refined when these factors recur. The decision model draws upon information produced by the cognitive and normative model, and helps to define the available choices and actions. The predictive model is typically used to simulate particular scenarios. It can be used to characterize the spatial and temporal extension of phenomena and events. Normal production flows and accidents (including effects such as the severity of a gas leak, and the concentration and diffusion of a toxic substance in the atmosphere) can be simulated in different contexts.

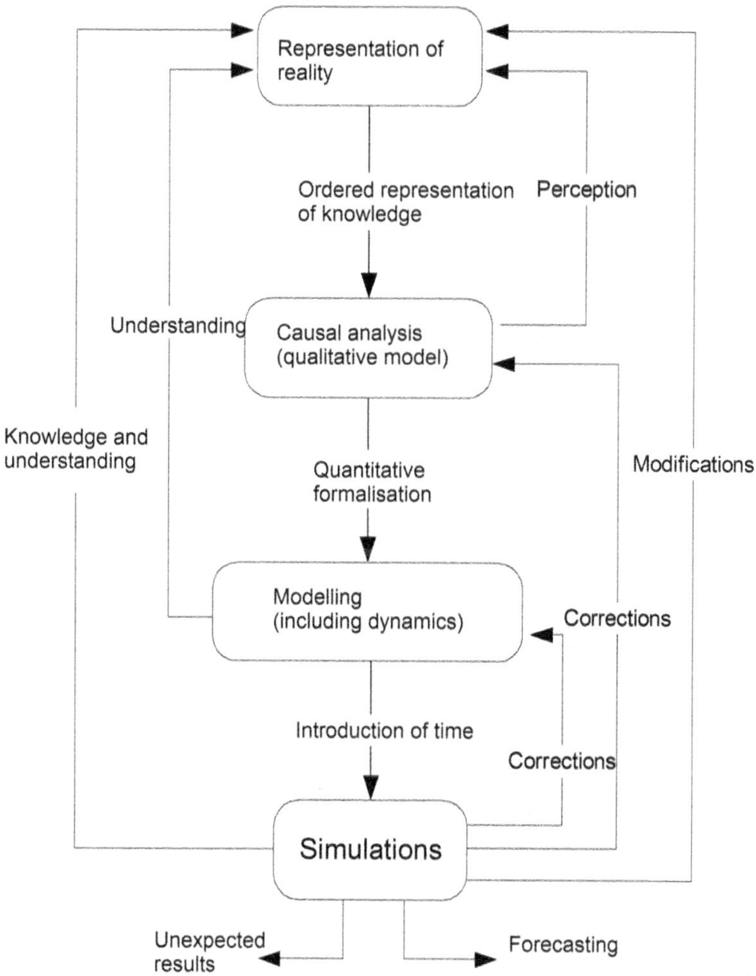

Fig. 1.2 The construction of a simulation model. (Translated from Donnadieu and Karsky 2002, p. 120)

These interrelated categories of models result in relatively elaborate system models, such as those developed for simulation. The foundations for these models are typically found elsewhere, and expressed formally in ways that are understandable and can be shared. The construction of a simulation model iterates between a representation of reality, which may initially be very simple or shared, and simulations that are the results of using a model that is based on the identification of causal phenomena (Fig. 1.2).

The dynamic risk analysis approach (presented in Chap. 2) is inspired by this system modelling approach and the simulation of system behaviour. The proposed simulation model should not be constructed on the assumption of the production of a stable, permanent tool that is defined once and for all, and which is used (unmodified) for forecasting when the need arises. The purpose of a simulation model is linked to the needs of a given project at a particular time. As time passes, the actual system is modified as a result of the lessons that are drawn from the model and applied to the organization and its procedures. This makes the model out of date and potentially inaccurate. Modelling is a process of creating and updating a model that satisfies the current needs and constraints. A *good* model should therefore be considered as one that is based on a specific methodology: it is reproducible and stable, adaptable and robust, and can be revised based on the results of the model simulation.

The dynamics of the model can become complex and counterintuitive as the number of system elements and their interactions increase. This makes it harder to determine the probable evolution of the system under particular conditions. Systemic modelling, which aims to produce simulations, addresses the system properties that are the basis of this complexity. The use of modelling methods makes it easier to understand systems, through the transcription of systemic phenomena from their expression as theoretical concepts to practical modelling tools. The rest of this book focuses on the system dynamics method.

1.2.2 System Dynamics: A Modelling and Simulation Method

In his *Industrial Dynamics* book, Forrester (1961) defines system dynamics as "a way of studying the behavior of industrial systems to show how policies, decisions, structures, and delays are interrelated to influence growth and stability." This requires the identification (through a graphical representation) of causal chains that define the evolution over time of a given system's components. These causal chains form feedback loops, which identify the processes that underlie counterintuitive phenomena using computer simulations. The graphical representation of feedback loops between elements of the system is called a *causal graph*. Most software packages used for developing simulation models provide a visual representation, in the form of a graph or a stock-flow diagram, which defines the system's elements as a function of their causal interactions. One such software package, STELLA®, is discussed in Chap. 2.

Forrester (1961) explains the detailed procedure for constructing models, which consists of five main steps:

- identify the key variables that continuously describe the status of the system's components in interaction;
- develop a hypothesis that may explain the dynamic behaviour, with a view to the proposed formalization of the system that distinguishes it from its environment;
- develop a formal model of the causal relationships between variables, based on knowledge and hypotheses, and identify feedback loops;

- decompose the causal phenomena, using a top-down approach, to understand complexity at the point where it affects dynamics;
- present these relationships in the form of differential equations in a mathematical framework that can be used in simulations.

System dynamics provides a way to study, model and simulate the processes that lead to changes in, or the maintenance of natural and artificial systems over time. These changes (if they occur) are observed when digital or ordered values are used as signals in the system being studied.

Forrester's method has been widely adopted, and has inspired many prospective models. Perhaps the most famous of these is found in the Meadows Report for the Club of Rome, entitled *Dynamics of Growth in a Finite World* (1974). The report's conclusions forecast the collapse of the world's population due to declining natural resources, increased pollution levels in the twentieth century and the depletion of non-renewable energy resources. The report educated industrialists and governments about the risks of uncontrolled growth.

The variables used in the Meadows Report relate to the demographics of the global human population, the overall level of pollution, the use of natural resources, the extent of farmland, etc. The report influenced the idea of sustainable development that was introduced in the late 1980s in the report of the World Commission on Environment and Development. A new edition of the Meadows Report was published in 2004, which confirmed some of the results from the original version.

The Meadows Report showed that the world's population could be stabilised through: birth control; the protection of resources and nature; limiting industrial emissions; and the conservation of arable land. The application of system dynamics to the study of the dynamics of the global population has led to the formalization of a model (World 3) that can represent the most significant variables. This approach can also be used to simulate the system's behaviour and explore different scenarios by varying its parameters. Some of these variations are political choices: the approach therefore contributes to the implementation of a decision support framework.

The application of system dynamics in industrial safety is designed to provide a formal decision support framework for industrialists. This framework is specific to the implementation of prevention and protection measures for workers, facilities, local residents and the environment.

Systems theory has also been combined with control theory to create the Systems-Theoretic Accident Model and Process (STAMP; Leveson 2004a, b). Unlike traditional accident models, which consider that accidents are the result of a chain of failures, the STAMP model adopts a systemic view of accidents (Leveson et al. 2003; Leveson 2004a, b; Leveson and Dulac, 2005; Stringfellow Herring et al. 2007). Industrial systems that have been studied using the STAMP method are based on a dynamic process that seeks to continuously adapt, by responding to internal and external changes to the system, in order to achieve their objectives. Safety is therefore considered as an emergent property of a system, and becomes a control problem that can be solved by strengthening safety constraints. These constraints do not apply directly to the system's elements but are imposed in the context of rules that define the interaction *between* these elements. Consequently, in the STAMP model, accidents result from a lack of control, or the improper application of safety constraints during the system's design, development and operational phases. From a systemic point of view, accidents are therefore the result of poor feedback that fails to provide a level of safety that corresponds to the optimal performance of the industrial system (an inappropriate or the lack of a recovery loop).

The basic concepts used in systems dynamics are described below. These provide the foundations for the work described above, and for the study of the industrial system that appears in Chap. 2.

1.2.2.1 Systemic Concepts in System Dynamics

System dynamics can be characterised using the following concepts (Donnadieu and Karsky 2002):

- state and force
- feedback loops
- nonlinearity
- deadlines; and
- structural change

These concepts are illustrated below, using a case study of a chlorine unloading unit that is used in the synthesis of plastics.

State and Force

A dynamic system can be understood in terms of its past and expected future changes. These changes are evaluated through a comparison of (null or actual) values at two points in time; these values continuously record the state of system elements. Changes in state are the result of interactions between elements. The direction of change in the value of a particular element depends on the values of the other elements with which it has a direct relationship. If the values of those other elements cancel each other out, the element does not change state; if there is a net change in values (non-null), the element changes state. These change values are referred to as *forces*.

The state of a chlorine stock (Cl_2) at the current time depends on the state at a particular previous point in time, and the forces—chlorine feed and consumption—that act on the state between the two time instants. This concept is reflected in the term *stock–flow*, which indicates this view of the state and the forces at play in a modelled system.

The relationship between an element and the other elements that act upon it via different forces is, from the systemic viewpoint, a circular causality. It is represented by a feedback loop as a function of the principle of the interaction.

Feedback Loops

Feedback occurs where the relationship between elements *A* and *B* influences the nature or intensity of the action of *B* on *A*. If the chlorine stock (element *A*) is empty, this state determines the value of the information that the stock manager sends to the chlorine supplier. The supplier responds by delivering the requested amount. If the chlorine stock is full and does not decrease (i.e., there is no consumption), the chlorine stock manager does not contact the supplier (i.e., the value of information is null), and the chlorine supplier does not make a delivery. If the supplier does not make a delivery when stocks are low, the stock manager renews their request: in this case, a failure to act (null value action) is the origin of another action with a non-null value, a request.

The thermostat is an example of an automated feedback loop. A desired state (temperature) is compared to the current temperature in a particular location, such as a room in a house. The difference between the two values (desired and actual) is information that determines whether to activate the heating system to reduce that difference and thereby achieve the desired temperature. Information feedback and action in the thermostat set the system dynamics, warming or cooling as appropriate. External conditions (the outdoor temperature) act on the system, but the reverse is not true. The outdoor temperature affects the indoor temperature (and the way that it changes): if it is hot outdoors, it is generally hot indoors too. The indoor temperature (and the way it changes), however, has little or no effect on the outdoor temperature.

There are two types of feedback loops. *Positive* feedback loops that amplify, strengthen, or are explosive; while *negative* feedback loops stabilize or control:

- Positive feedback loop: the set of causal relationships that establish feedback on a system element define a force that changes the value of that element. This value, in stable conditions of causal relations, is constantly changing to a value that is increasingly different to the initial value. If the force of change is proportional to the value of the element, this force acts exponentially over time, and the loop is termed an *explosive* loop.
- Negative feedback loop: the set of causal relationships that establish feedback on a system element define a force that maintains the value of that element. This type of loop leads to a situation of equilibrium or stability.

These feedback loops form part of the system structure. The number of feedback loops in a modelled system depends on its spatial dimensions (i.e., the number of items and the number of relationships between them). These loops are the results of one-way causal relationships between elements that define the reflexivity of each element in the feedback loop. One-way causal relations (a change in element *A* is due to a change in element *B*) are described as positive, or negative, depending on the direction of the change of the two elements. The causality is *positive* when a rise in *A* is due to a rise in *B*, or when *A* falls due to a fall in *B*: the direction of change is the same, and the dynamics are qualitatively similar. The causality is negative when a rise in *A* is due to a fall in B, or when a fall in A is due to a rise in B: the direction of change is different, and the dynamic is qualitatively different.

In a feedback loop, the number of positive and negative causalities that follow each other determines whether the entire feedback loop is positive (amplifying) or negative (stabilizing):

- A feedback loop is positive (reinforcement) when it only has positive relationships or has an even number of negative relationships;
- A feedback loop is negative (maintaining) when it has an odd number of negative relationships.

This can be compared to cogwheels in a gearbox, where the direction in which the last wheel turns—and whether it is the same as that of the first wheel or not—depends solely on the number of wheels.

Generally, causal properties are represented by a flow diagram that shows the names of the interacting elements linked by arrows that indicate the direction of the causality between the two elements of the loop. The type of causality (positive or negative) is indicated by a "+" or a "–" at the end of, or alongside the arrow. The terms *positive* and *negative* can be confusing: positive can be simply interpreted as meaning an increase; negative as a decrease. For this reason, the term *same* (designated by an "S" rather than a "+") may be used instead of *positive*, and *opposite* ("O" rather than a "–") may be used instead of *negative*; the terms *similar* and *opposed* are also sometimes used.

The arrangement and the nature of the feedback loops, i.e., the causal structure of the modelled system, can be used to determine its characteristic dynamics. There are several archetypal causal structures. Figure 1.3, for example, shows the archetype of the feedback loop that creates a dynamic equilibrium. The current state of an element, such as the level of chlorine stock, is compared with the desired stock level (e.g., maximum stock). The difference between the two levels is determined by a negative causal relationship from the present level: when the stock falls, the difference rises; when the stock rises, the difference falls. The dynamics reflect an opposition in the changes in the values of the chlorine stock level.

The difference between the levels is a piece of information that is the cause of an action: an increase (or not) in stock. The greater the difference, the more action will be needed to increase the stock level to reach the desired level (i.e., until the difference is reduced to zero). The causal relationship establishes a similar dynamic of the two terms, therefore it is positive. The action (delivering chlorine) finally changes the state (increases the level of the stock).

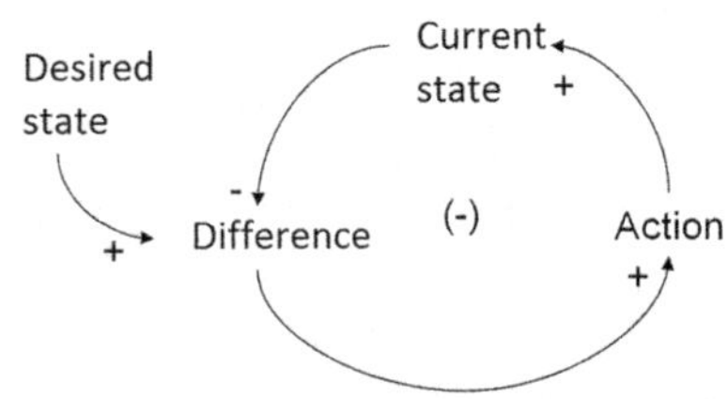

Fig. 1.3 Archetypal causal structure of a feedback loop

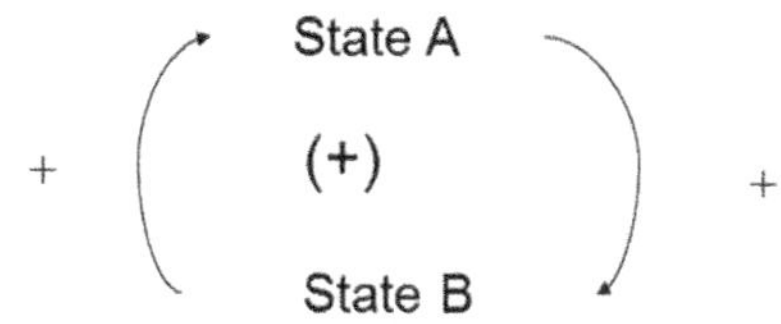

Fig. 1.4 Archetypal causal structure of a reinforcing loop

Similarly, if the desired level at a given time is less than the current level, then the action will consist of consuming chlorine to reach the desired level. The current level evolves in the same direction as the action. The feedback loop that acts from the state of the current element (stock) on itself, via a difference in elements with a desired state and an action has three links, one negative and two positive. In other words, when the stock value falls, the difference increases, and when the difference increases, the action increases, when the action increases, the stock increases. The current state (the stock level), therefore receives an action that is the reverse of its initial dynamic, leading to greater stabilization and balance, converging on the value of the desired state. A symbol is usually placed in the centre of the feedback loop to identify this, here, a "–" to indicate a negative loop, or opposition.

A positive feedback loop can be illustrated by the simple archetype of reinforcement or a runaway system (Fig. 1.4): a state *A* acts on a state *B* causing a change in the same direction. In turn, the state of *B* affects the state *A* causing a change in the same direction. If *A* increases, then *B* increases, *A* increases again, etc. If *A* decreases, *B* also decreases, and *A* decreases again. This feedback loop is denoted by a "+" that qualifies its dynamic property.

These simple dynamics are representative of systems with relatively simple causal structures. The dynamics shown above (Fig. 1.5) relate to systems with a single feedback loop. Most natural systems and those that are organized by humans (both social and technical) have multiple feedback loops, and some of these may act on the same element. The feedback loops can have conflicting or complementary properties, and affect an element with forces that are difficult to assess by a simple analysis. The forces that result from this complex causal combination generally define a non-linear dynamic, which underlines the importance of the need to use simulation to understand the system's behaviour.

Non-linearity

Non-linearity is typical of non-proportional causal relations. The evolution of a value in a causal chain can define a force that is not proportional to the state that engenders it, and this force itself acts in a non-proportional way on the element that is changed.

This non-linearity arises out of the characteristics of the system's elements, particularly their dynamic properties. Natural, adaptive and very sensitive systems exhibit non-linear behaviour that is more marked than it is in artificial systems, in which the elements have fairly simple properties (binary states or linear gradients).

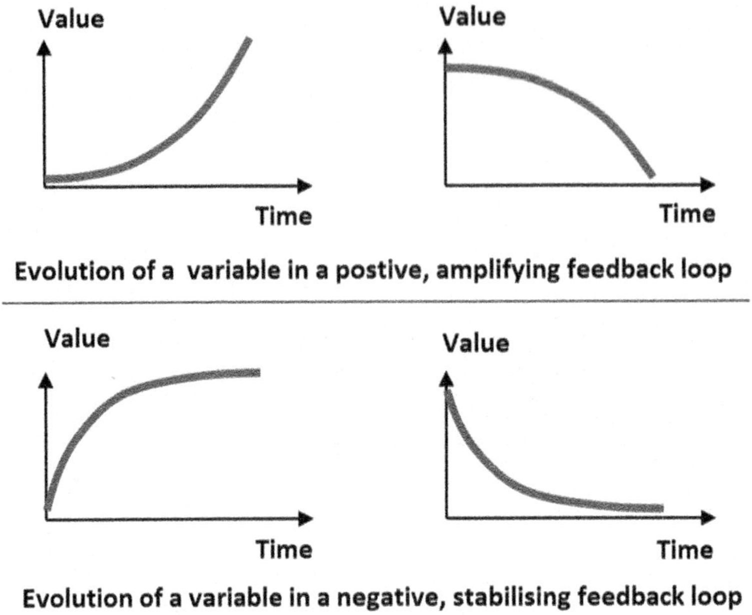

Fig. 1.5 Evolution of positive and negative feedback loops

Fig. 1.6 Archetypal causal structure of the need for change

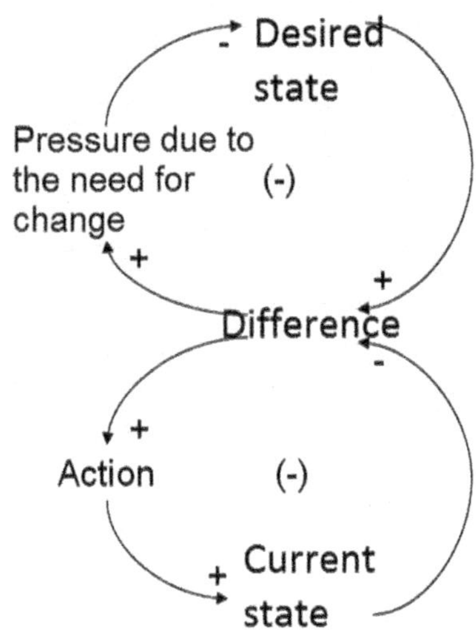

The non-linearity in artificial systems is usually determined by the number and variety of elements.

These properties can be illustrated using a system that is characterised by a causal structure made up of two negative feedback loops (Fig. 1.6), which highlight a need for change. Compared to Fig. 1.2, the archetypal equilibrium loop, the desired state changes as a function of the difference from the current state.

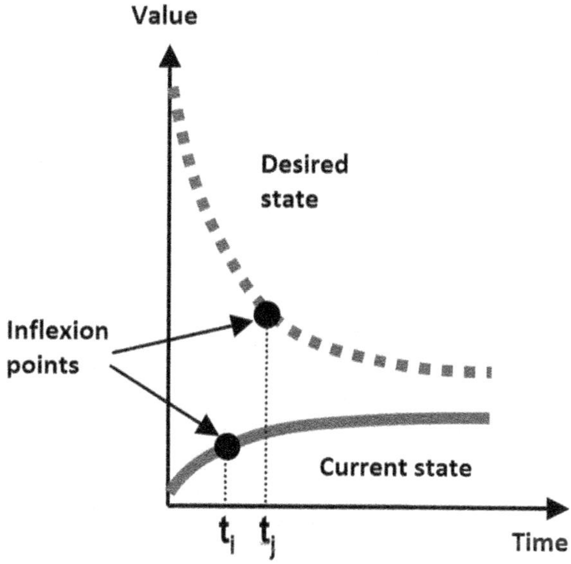

Fig. 1.7 Example of temporal variation in the changing need

The desired state is adaptive, relative to the current state, in order to minimize the difference at the heart of the structure. The change in the difference triggers a change force on the need for change, setting a value for the desired state. This reduces the difference with respect to the current state, on which the action of the search for an equilibrium is least. Depending on the properties of the change force and the action, the reduction in the difference between the desired state and the current state can vary disproportionately.

Moreover, the change force can, for example, change the desired state more slowly than the action changes the current state (Fig. 1.7). This creates variations that are shifted in time, indicating a characteristic asynchronous dynamic.

Another archetypal example of nonlinearity is the growth limit. This involves two feedback loops in the causal structure: one is positive, and provides the basis for growth; the other is negative and defines the growth limit (Fig. 1.8). The current state evolves over time through the action of growth: the higher the value of the current state, the more the action of growth increases. This loop is therefore exponential. The current state's growth, through the positive loop, is limited by a negative loop that is related to a limiting state, i.e., a maximum value of the state. The higher the value of the state, the more the slowing action increases which, in turn, brings about a reduction in the value of the current state.

At some point the slowing action becomes more powerful than the growing action, and the value of the state approaches the asymptote of the limiting state. The shape of this logistic curve resembles an "S" (Fig. 1.9).

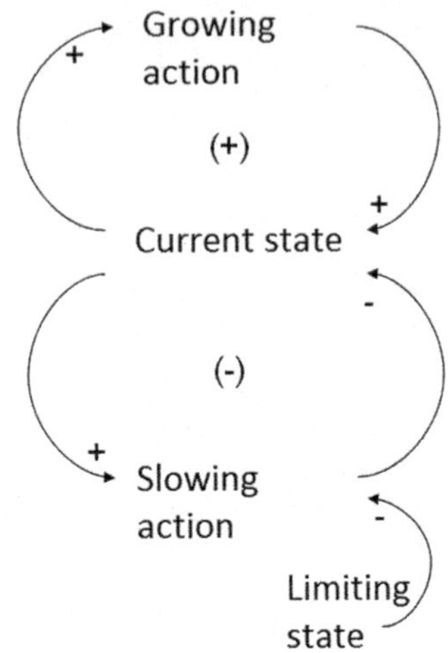

Fig. 1.8 Causal structure archetype of growth limit

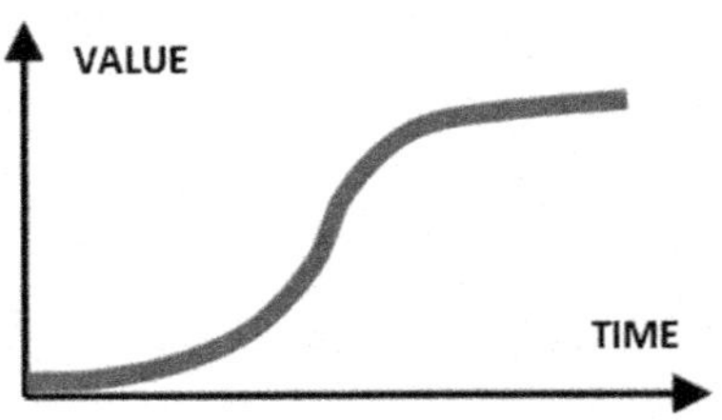

Fig. 1.9 Logistic-type variation

Deadlines

Causality indicates a relationship between two elements, which results in one of them changing. The change takes place over a time interval that indicates when the resulting value is achieved. The effect of the change is seen after the point in time when the cause happened: the propagation of causal phenomena in feedback loops are marked by time delays. If a causal system incorporates several feedback loops, each may have a different time delay. A dynamic that is initially marked by a particular trend can change form, beginning at the moment when a cause marked by a deadline is taken into account by the element that experiences the effect. The impact of a time delay on the dynamics of a system can be illustrated by the archetype of the unstable load (Fig. 1.10).

Using the example of the unloading and distribution of chlorine, any problems may be detected by their symptoms. According to the information that is available to solve the problem and thus remove the symptom, different solutions can be envisaged. These solutions may require resources that take different lengths of time to implement. Depending on the application of these solutions, some will have a short-term impact, so the symptoms will re-appear, i.e., the problem will not be fully

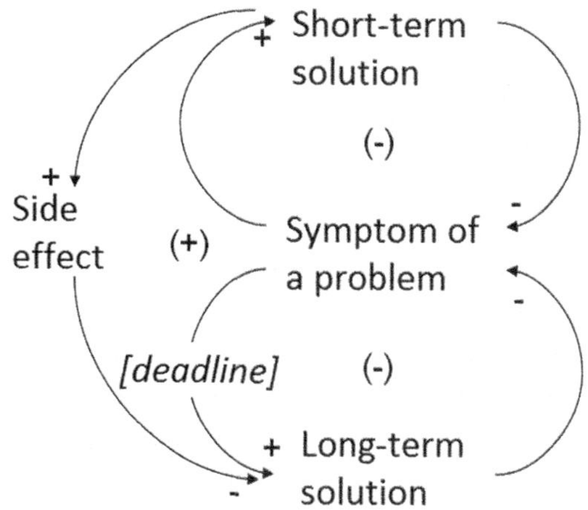

Fig. 1.10 Causal structure archetype of the unstable load

resolved; other solutions will create conditions that make the problem *and* its symptoms disappear. In a single causal structure, it is possible to indicate the impact of the importance of the deadline for the implementation of the solution in the long term, potentially leading to its ineffectiveness if, in the meantime, the short-term solution has amplified the problem.

Structural Change

The notion of structural change differs between theory and practice. From a theoretical point of view, a system has a structure that undergoes changes from the perspective of causalities that can arise at any point in time. These variations reflect the history of the system. When modelling a system and its structure, consideration has to be given not only to the causalities that are detectable or measurable by past and present observation, but also to the potential causalities—possible, probable and certain—that were not active when the model was designed.

The systems literature distinguishes so-called deterministic systems from stochastic systems, i.e., those characterised by random events. In these stochastic systems, inactive potential causalities can act at any time, while causalities observed at any given point in time may stop randomly. The difference between determinism and randomness is partly related to the degree of complexity of the system, and to the degree of knowledge of this complexity. In complex systems, such as populations of interacting individuals, structural changes can be particularly marked by bifurcations. A dynamic, starting from a small change in the energy or information in the system, can lead to a significant qualitative change.

In Darwin's theory of the evolution of living species, the brutal conditions of isolation of individuals from the same population explains the differentiated qualitative evolution in the resulting two populations: one experiences environmental and competition conditions, while the other does not. This results in a bifurcation in the dynamics of the system, and the potential emergence of two different species over a long period of time. Some systems modelling methods are better than others at highlighting this emergence of new structures.

1.2.2.2 Principle Characteristics of System Dynamics Models

In system dynamics, the identification of positive and negative feedback loops is the point when the qualitative expertise in the analysed system is made explicit. The quantitative formalization that translates this qualitative knowledge into a simulation model follows several stages, as described below.

- Use of the concept of causal structure. In the systemic approach, the structure corresponds to the set of sustainable relationships between the system's elements. An element is part of a system only if it is related to other elements: this is the relationship that provides the foundations of the system. The dynamic operation of a system is thus one of cause and effect relations between its elements, governed by feedback loops.
- The mathematical formalization of state variables using differential equations. The change in value of a state variable in any given time period depends on the balance between the values of the input and output variables associated with it, and its own value in the previous time period. This can be expressed as a differential equation:

$$\frac{dX}{dt} = W - Z$$

Where X is the state variable, W is the input, Z is the output, and t is time.

Modelling the system consists of writing the set of differential equations that expresses the values of state variables, and hence the system state, at any point in time. The sequence of states over time corresponds to the trajectory of the modelled system, and makes it possible to evaluate the system's dynamic behaviour.

- Programming a computer simulation. Solving a large set of differential equations requires a software package or computer programme. These packages can be used to develop simulation scenarios and to analyse the model's sensitivity to different parameters and initial conditions. The results of these simulations constitute a type of experiment that is typically difficult or even impossible, to implement when it takes the form of, for example, complex and multi-sectoral or multi-actor systems.

These three stages are encapsulated in the STELLA® simulation and modelling tool, described in Chap. 2, which has been applied to risk management in an industrial plant for the unloading and supply of chlorine. The next point addresses the various contributions of system dynamics in the context of industrial activities.

1.2.3 The Contribution of System Dynamics to Industry

Industry has a longstanding interest in systems thinking, particularly around dynamic systems, because it can be used to describe the behaviour of an industrial unit, a factory or a process and the relationships between system's parameters. One of the reasons why industry views the systemic approach as essential is that it is rooted in the worlds of science and engineering. Industry has also made a major contribution to the development of the systemic approach through its needs and modes of organization.

System dynamics can be used to analyse industrial activities. Economic activities—design, production, logistics related to the supply of raw materials and energy sources and their distribution—are all subject to enormous problems linked to deadlines and delays, competition, and physical constraints on the social organization of work. These are all topics that provide a very rich field for the application of the systemic approach, and particularly modelling and simulation. The systemic approach is especially useful because it is considered both as a business tool and as a production tool.

An industrial company—as an economic actor—can be considered as a system, whose interacting elements are designed to produce goods in conditions that are *a priori* economically viable. The boundaries of this system vary according to the question to be investigated and the problems to be solved: all of the raw materials that must be transformed, and the sources of energy needed to achieve this transformation can be incorporated into the industrial system. One example is the type of work that places industry at the interface between society and the factors that relate to human and social sciences (economics, psychology, sociology, ergonomics, etc.). These different domains interact over time, and are seen in the spatial heterogeneity of different administrative levels (e.g., local, regional, national). An industrial plant incorporates interactions that may be considered beneficial (e.g., creation of new jobs) or harmful (e.g., environmental pollution). As the environment and its component elements change, these interactions produce a changing dynamic that needs to be controlled.

Industrial production structures are often modelled in order to identify problems and find solutions that increase productivity. Just-in-time or lean methods, for example, can be modelled in order to understand both potential bottlenecks in the system *and* opportunities for improvement. These methods are particularly useful in the capital goods sector, which is dominated by supply and demand cycles.

In the process industries, the transformation of raw materials into commodities (chemicals, petrochemicals, etc.) is energy-intensive. This means that, in addition to economic risks, other major risks (e.g., to people and to the environment) arising out of energy usage need to be considered.

To illustrate the contribution of system dynamics modelling in industry, several studies that have applied Forrester's (1961) concepts will be used. These examples, which are discussed below, cover: supply chains; risk management in companies and the environment; and the emergence of accidents.

1.2.3.1 Application to Supply Chains

System dynamics modelling was used by Pierreval et al. (2007) to develop a supply chain model for a large company in the automotive industry. The aim was to evaluate long-term trends in the dynamic behaviour of car production facilities. The organization is characterized by a network of subcontractors, and has to cope with extreme changes in demand for some components.

Pierreval et al.'s (*op cit.*) model represents the overall behaviour of a set of factories working in an integrated supply chain, based on decisions that will be applied to those factories in the next three to six months. Each production facility is divided into *logistical units*. A logistical unit is defined as "a physical sector of a factory, which can either be in the same factory or an external supplier, which has a delivery zone for the objects or components that are produced, and the human and technical resources that are needed to achieve production targets." The decision to represent the logistical units defines the level of the real system that is represented by the model. It is a relatively macroscopic level, which focuses on the overall behaviour of the system.

They followed Forrester's principles to develop their model, and identified five logistical units. These units were characterised by: their activity (assembly, production); their location; and the types of objects they produced. These units have to adhere to schedules that specify production rates (units per day). In broad terms, the supply chain follows the just-in-time or lean philosophy.

Their simulations focused on the study of the dynamic behaviour of the inventory over a four-month period. The aim was to identify production shortages that are weighted as input or output flows for each unit. The graphical results show the quantity of elements produced and the quantities that are stored on a daily basis in each logistical unit. The dynamic is characterised by (normal) weekly oscillations reflecting the plant's closure on Sunday. The identification of stock shortages at certain units indicates that they can accommodate an increase in input flow elements, thereby increasing their stock levels. A shortage at any point in the supply chain usually has repercussions for the whole chain, both upstream and downstream. The model allows turnaround times in each logistic unit to be examined, and the actual and potential production levels to be compared. The model can be used in different scenarios to identify those units that need to change their production or storage capacity in order to avoid creating shortages in the supply chain.

1.2.3.2 Application to Risk Management and Environmental Impact

In the field of industrial safety, systemic modelling is generally used to identify the areas where delays or changes that are hard to detect manually or using automatic sensors can lead to unsafe situations and, potentially, to accidents. The value of modelling here lies at the interface between the technical and the social (human) organizations. Different scenarios can be investigated using the same model to explore the different situations arising from small or large failures. For example, in an industrial process, worn parts may lead to increased energy consumption over time, up to a point where the levels are unsafe, and an accident can occur. Similarly, the social organization, (e.g., operators and supervisors) may, under stressful conditions, fail to correctly handle an abnormal situation, and ultimately trigger an accident. In this case, modelling is seen as a *soft* method. It makes it possible to experiment with situations that are very demanding in practice. The investigation of critical situations is particularly difficult to achieve on a real-life scale, not only because of the direct risks, but also because of the economic costs of a failure.

Perhaps one of the most interesting contributions of the systemic approach to industry is the implementation of Safety Management Systems (SMS; Santos-Reyes and Beard 2001, 2008). Since the European Seveso II Directive was passed into French law in 1999, classified installations have been required to implement an SMS to "help to prevent and control hazards associated with major accidents and to limit the consequences" (Official Journal of the European Communities, 1996). The SMS must provide clear evidence that the elements that have been put in place to support a policy for the prevention of major accidents and its application have been fully integrated into the company. These elements include: a description of the roles and the responsibilities of staff; the identification and analysis of major risks; the definition of control systems in normal system operation and during maintenance operations; change management procedures; emergency planning, and so on.

Santos-Reyes and Beard explored how the systemic approach could be used to define a systemic SMS (SSMS) that takes into account the principal safety features that prevent the occurrence of major accidents. The main contribution of their work was to have a real SMS that could guarantee its prevention objectives, taking into account the relations between different safety features and their dynamics over time. This SSMS is built by decomposing the system into eight sub-systems covering the following safety functions:

- The implementation of regulations applicable to industrial activities and associated risks;
- Coordination of safety operations with the company's activities and its environment;
- The definition of how safety is organized in the company;
- Safety system audits;
- Continuous improvement of the safety system, notably through support for research and development and the integration of the results;

- Ensuring the confidentiality of information provided by the system used by employees to report problems that they encounter in the course of their work;
- Monitoring changes to safety regulations and the decisions taken by the company in response;
- The establishment of a dedicated communication hotline between the relevant stakeholders in case of emergency.

Modelling an SMS is particularly interesting because it focuses on the relationships between the company's safety functions. This approach has not been applied to a case study and, moreover, is not based on a tool that allows the modelling of the operation of the system from a dynamic point of view.

The STELLA® software package has been used to investigate the consequences of industrial activities on the environment and the ecology of living beings. For example, it has been used to explore the use of plant species for the remediation of contaminated soil by 1,4 dioxane and Tri-Ammonium Toluene (Ouyang 2002; Ouyang et al. 2007). The roots, stem and leaves of the Italian poplar (*Populus nigra* var. *Italica* Duroi = *Populus fastigiata* Foug.) were modelled. STELLA® was used to simulate transfers of pollutants in the three parts of the plant in order to estimate the accumulation and changes in the tree's metabolism over time. In this way it was possible to assess the effectiveness of soil remediation strategies at industrial sites, and to estimate how long the process would take for different plant species.

STELLA® has also been used to model the impact of industrial activities on ecosystems, focusing in particular on the coupling between industrial and natural systems (Reap 2004). This work is partly based on the University of Maryland Institute for Ecological Economics' modelling of natural and economic systems and the relation between them (Costanza and Gottlieb 1998; Costanza et al. 2002; Boumans et al. 2002). The objective is to evaluate the impact of industrial activities on the environment in the context of sustainable water resources and biomass management. This coupling is represented in STELLA® by developing a model that provides a spatial view of the results of the simulation of a toxic release in a particular territory (Hunting Creek, Maryland, USA). The area is divided into discrete pixels that are classified according to their use (forest, agriculture, water bodies, etc.) to show the relationship between industrial activities and the environment. The results of the simulations provide a basis for supporting decisions about land management in the context of industrial waste management.

1.2.3.3 Application to the Understanding of Industrial Accidents

After studying the conditions that lead to incidents and accidents Cooke and Rohleder (2006) developed a systemic learning model for incidents with the aim of preventing future critical situations. The model assumes that the lessons learned from incidents are poorly integrated into existing monitoring procedures, which are consequently deemed inappropriate and marked by a failure to capture past experience.

Following the principles of dynamic systemic modelling, Cooke and Rohleder identified several feedback loops between sub-systems for production, safety and learning from incidents. Figures 1.11, 1.12 and 1.13 illustrate the principles of systemic modelling discussed above.

The first feedback loop identifies a causal link between actual and expected productivity (Fig. 1.11). This is a negative feedback loop: actual productivity tends to reduce the gap with productivity targets, albeit with a delay. This is an example of the balanced loop causal structure archetype.

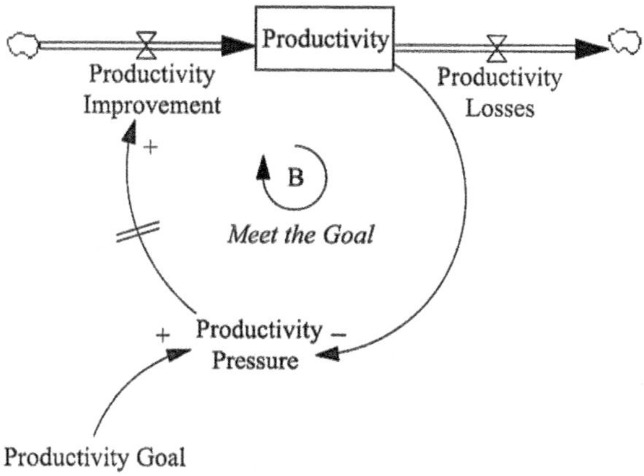

Fig. 1.11 System for the organization of productivity. (Cooke and Rohleder 2006, p. 220)

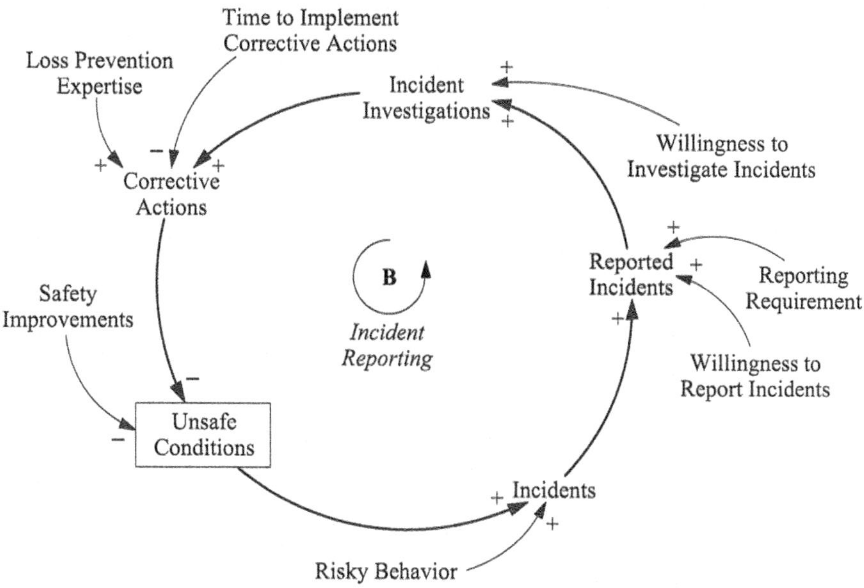

Fig. 1.12 The incident tracking system. (Cooke and Rohleder 2006, p. 222)

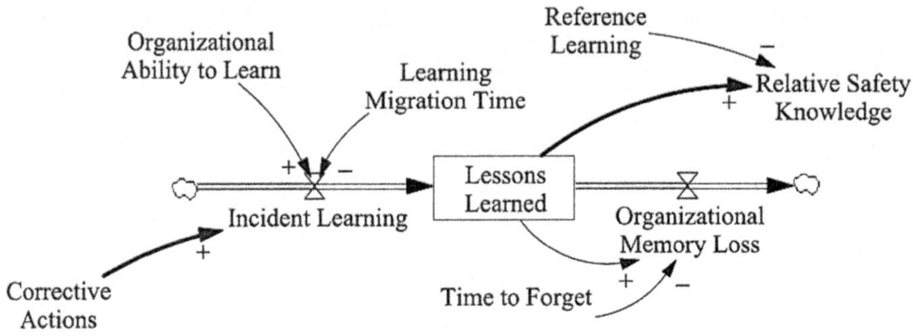

Fig. 1.13 The stock-flow structure of learning from incidents. (Cooke and Rohleder 2006, p. 226).

There is a delay between the pressure to produce, and the effect of productivity goals, which is modified by actual productivity.

The learning from incidents subsystem is also a negative feedback loop (Fig. 1.12) that, after an incident, leads to an investigation and corrective action that changes an unsafe state to a safe one, via a reporting mechanism. This loop incorporates the influence of both quantitative and qualitative elements. The quantitative elements, such as the time to implement corrective actions, can be evaluated objectively; the qualitative elements, such as the willingness to investigate incidents and loss prevention expertise have to be evaluated subjectively by an expert.

Other sub-systems that add complexity to the general system are also included in the model. It should be noted that the stock-flow diagram (Fig. 1.13) assumes that the proposed representation of the structure of learning from incidents can be implemented in software. This structure clearly demonstrates the importance of elapsed time and forgetting as a potential cause of risk in future situations.

The lessons that are learned from an incident are considered as a stock which grows as a function of an inflow (learning from incidents) and an outflow (loss of organizational memory). The conditions for learning and forgetting can change, depending on time constraints and capacity. This qualitative representation and its subsequent implementation in software, can be used to represent variations between gains and loss of experience using quantitative levels. Simulating extended time periods (ten years or more), makes it possible to assess which system elements appear to be strategic and hence should be maintained or improved in order to optimize the safety information system.

The system dynamics modelling approach has also been used to examine other accidents and safety situations, including:

- The 1992 Westray mining explosion (Cooke and Rohleder 2006);
- The 1998 Piper Alpha oil rig fire (Paté-Cornell 1993);
- The Walkerton water contamination accident (Leveson et al. 2003);
- Safety in a nuclear power plant (Jaekook et al. 2004);
- Dynamic reliability of a nuclear power plant (Kyung and Moosung 2005); and
- Risk analysis in the context of industrial facilities governed by the Seveso directive (Garbolino et al. 2009).

The next chapter shows how dynamic modelling can be applied to a chlorine unloading facility in a plastics synthesis factory. This application draws heavily upon system dynamics concepts to construct and simulate the behaviour of the system. Various failure scenarios form the context for these simulations, which are then placed in a risk analysis context to develop a dynamic risk analysis approach.

References

Boulding K (1956) General systems theory: the skeleton of science. Manag Sci 2:197–208

Boumans R, Costanza R, Farley J, Wilson MA, Portela R, Rotmans J, Villa F, Grasso M, (2002) Modeling the dynamics of the integrated earth system and the value of global ecosystem services using the GUMBO model. Ecol Econ 41:Special Issue "The Dynamics and Value of Ecosystem Services: Integrating Economic and Ecological Perspectives", 529–560

Cooke DL, Rohleder TR (2006) Learning from incidents: from normal accidents to high reliability. Syst Dyn Rev 22(3):213–239

Costanza R, Gottlieb S (1998) Modelling ecological and economic systems with STELLA®: part II. Ecol Model 112(2-3):81–84

Costanza R, Voinov A, Boumans R, Maxwell T, Villa F, Wainger L, Voinov H (2002) Integrated Ecological Economic Modeling of the Patuxent River Watershed, Maryland. Ecol Monogr 72(2):203–231

De Rosnay J (1975) Le Macroscope. Vers une vision globale. Seuil, Paris, France

Descartes R 1637 (2014) Discourse on the method of rightly conducting the reason, and seeking truth in the sciences. Veitch J. (Translator). CreateSpace Independent Publishing Platform, 34 p

Donnadieu G, Karsky M (2002) La systémique, penser et agir dans la complexité. Editions Liaisons, Paris, 269 p

Durand D (2006) La systémique. Editions PUF, Collection Que sais-je? 127 p

Forrester JW (1961). Industrial dynamics. MIT Press, Cambridge, pp 464

Garbolino E, Chery JP, Guarnieri F (2009) Dynamic systems modelling to improve risk analysis in the context of seveso industries. Chem Eng Trans 17:373–378

Jaekook Y, Namsung A, Moosung JA (2004) A quantitative assessment of organizational factors affecting safety using system dynamics model. J Kor Nucl Soc 36(1):64–72

Kyung MK, Moosung J (2005) A quantitative assessment of LCOs for operations using system dynamics. Reliab Eng Syst Saf 87(2):211–222

Le Moigne J-L (1977) La théorie du système général. Editions PUF, Collection Systèmes-Décisions, 258 p

Le Moigne J-L (1983) La théorie du système général. Théorie de la modélisation. Editions PUF, Collection Systèmes-Décisions, deuxième édition, 320 p

Le Moigne J-L (1990) La modélisation des systèmes complexes. Editions Dunod, Collection Afcet-Systèmes, 178 p

Leveson N (2004a) A new accident model for engineering safer systems. Saf Sci 42(4):237–270

Leveson NG (2004b) The role of software in spacecraft accidents. J Spacecr Rocket 41(4):564–575

Leveson N, Dulac N (2005) Safety and risk driven design in complex systems of systems. 1st NASA/AIAA space exploration conference, Orlando, February 2005

Leveson N, Daouk M, Dulac N, Marais K (2003) A systems Theoritic approach to safety engineering , October 30. Massachusetts Institute of Technology, Cambridge, 28 p

Meadows DH, Randers J, Meadows DL (2004) Limits to growth: the 30-Year update. 3rd edn, Chelsea Green Publishing, London, 338 p

Morin E (2005) Introduction à la pensée complexe. Editions du Seuil, collection Points, 158 p

Ouyang Y (2002) Phytoremediation: modeling plant uptake and contaminant transport in the soil–plant–atmosphere continuum. J Hydrol 266:66–82

Ouyang Y, Huang HC, Huang YD, Lin D, Cui L (2007) Simulating uptake and transport of TNT by plants using STELLA®. Chemosphere 69:1245–1252

Pascal B, 1670 (2012) The thoughts of blaise pascal. Boer PA (ed), Kegan CP (Translator). CreateSpace Independent Publishing Platform, Scotts Valley, 404 p

Paté-Cornell E (1993) Learning from the piper alpha accident: a postmortem analysis of technical and organizational factors. Risk Anal 13(2):215–232

Patrick GTW, Bywater I (2014) Heraclitus of Ephesus: the fragments of the work of Heraclitus of Ephesus on nature and Heracliti Ephesii Reliquiae. Literary Licensing, LLC, Whitefish, 244 p

Perilhon P (2003) MOSAR: Présentation de la méthode. Techniques de l'Ingénieur, SE 4 060, 16 p

Pierreval H, Bruniaux R, Caux C (2007) A continuous simulation approach for supply chains in the automotive industry. Simul Model Pract Theory 15:185–198

Reap JJ (2004) Plants in the garden: an approach to modelling the impact of industrial activities in ecosystems. Thesis presented to the Georgian Institute of Technology, 195 p

Santos-Reyes J, Beard AN (2001) A systemic approach to fire safety management. Fire Saf J 36:359–390

Santos-Reyes J, Beard AN (2008) A systemic approach to managing safety. J Loss Prev Process Ind 21:15–28

Stringfellow Herring M, Owens BD, Leveson N, Ingham M, Weiss KA (2007) A safety-driven, model-based system engineering methodology, part I. MIT technical report, December 2007, 56 p

von Bertalanffy L (1968) General system theory. Foundations, development, applications. Georges Braziller, New York

Zwirn HP (2006) Les systèmes complexes. Mathématiques et biologie. Editions Odile Jacob, Paris, 219 p

Chapter 2
Systems Dynamics Applied to the Analysis of Risk at an Industrial Installation

Emmanuel Garbolino, Jean-Pierre Chéry, and Franck Guarnieri

This chapter presents a framework for the implementation of the dynamic modelling of systems to support risk management at an industrial facility. This approach, implemented using the STELLA® software package, provides the decision maker with a way to model a system and simulate its behaviour over time. It takes a dynamic approach to the analysis of industrial risks, based on four complementary stages that facilitate detailed analyses and the continuous improvement of risk management (Fig. 2.1):

- **Design of the dynamic model and simulation of system behaviour**: This stage involves: identifying the variables that describe the state of continuously interacting system components; defining hypotheses that establish the interactions with a view to formalizing the proposed system; developing a model of causal relationships between variables; formulating these relationships as differential equations; and implementing them in software (Forrester 1961; Donnadieu and Karsky 2002).
- **Comprehensive failure analysis**: This stage uses a classical risk analysis method, HAZOP (Andrews and Moss 2002). This method is used to try to identify all potential failures. Using the dynamic model, potential changes in the system's behaviour can be analysed.
- **Comprehensive simulation of the consequences of failures**: This stage uses the PHAST® software package (see the DNV website) to simulate hazardous phenomena (toxic emissions, overpressure, thermal flows, etc.) and assess their potential consequences.

E. Garbolino (✉) · F. Guarnieri
MINES ParisTech/PSL Research University, CRC, Sophia Antipolis Cedex, France
e-mail: emmanuel.garbolino@mines-paristech.fr; guarnieri@mines-paristech.fr

J.-P. Chéry
AgroParisTech, Montpellier, France
e-mail: jean-pierre.chery@teledetection.fr

F. Guarnieri, E. Garbolino (eds.), *Safety Dynamics*, Advanced Sciences and Technologies for Security Applications,
https://doi.org/10.1007/978-3-319-96259-7_2

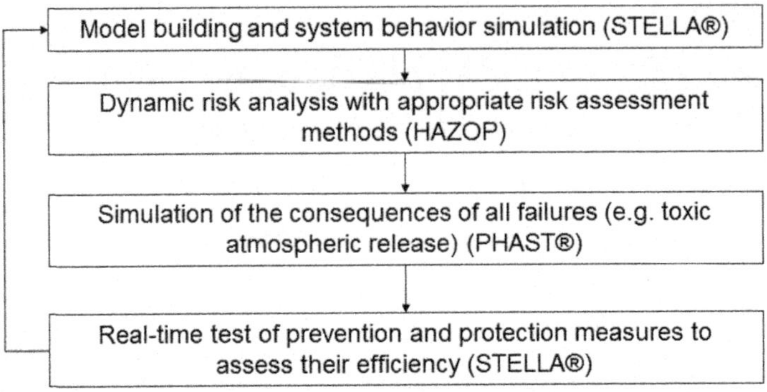

Fig. 2.1 General approach to the dynamic analysis of risk in an industrial facility

- **Testing of prevention and protection methods**: This stage consists of assessing the effectiveness of prevention, protection and backup measures. It makes it possible to define new measures, if necessary. The model can then be re-developed based on the assumption that the identified prevention and protection measures have been implemented (Garbolino et al. 2009).

This chapter introduces the industrial facility that is used as the case study and, in particular, the at-risk activity of the unloading and supply of chlorine. The failure analysis is based on a dynamic simulation of the unloading facility and the chlorine line in order to assess behaviour in the case of a malfunction. Specifically, it simulates leaks of liquid chlorine (in the unloading phase), and of chlorine gas (at the end of the supply line).

2.1 The Case Study and Risk Prevention Challenges

The company, a world leader in the manufacture of high-performance plastics, has production sites in several countries including France. The case study is based on a site that uses several hazardous substances (chlorine, xylene, ammonia, etc.) to synthesise plastics.

The hazardous nature of these chemicals (fire, explosion, toxicity) and the quantities stored *in situ,* means that the site is a *Classified Installation for Environmental Protection* (*Installation Classée pour la Protection de l'environnement* ICPE) and is subject to public scrutiny (*Autorisation avec Servitude*). The regulatory framework for the industry is laid down in French law and the European Union's Seveso directives. The site operator must conduct a hazard study every 5 years, and submit it to the appropriate public authorities, such as the Inspectors of Classified Installations (*Direction Régionale de l'Environnement, de l'Aménagement et du Logement*, DREAL) and local officials (notably the departmental Prefect). These representatives are responsible for the ongoing authorization of industrial activities.

The hazard study includes a risk analysis, and an assessment of the consequences of potential accident scenarios, taking into account the barriers implemented by the operator to reduce the occurrence and consequences of the identified hazards. The hazard study, which is the responsibility of the site operator, costs between 50,000 and 100,000 US dollars for a new installation, even when a third-party provides the necessary expertise.

Government inspectors provide a technical opinion on the hazard study and can request an expert third party to carry out more in-depth analyses of risk and accident scenarios. Given the magnitude of the risks involved, the operator is obliged to provide the best-quality risk analyses possible. The risk analysis is carried out by a working group, led by the Process Department, the Health, Safety, Quality and Environment (QHSE) manager and other staff with risk management responsibilities (process engineers, operators, maintenance engineers, etc.). Analyses require significant resources and can last several days depending on factors such as the complexity of the system, the procedures that are implemented, and how tasks are allocated.

Depending on the study's outcome, the operator may have to put in place emergency procedures, and implement an Internal Operations Plan (IOP). This plan is activated when an accident occurs on site. If the consequences of the accident are likely to extend beyond the site's perimeter, the operator and emergency services trigger the Specific Response Plan (SRP), which is authorised by the local Prefect. The SRP covers the most hazardous scenarios that may have consequences for the local population and the environment. These scenarios provide the authorities and civil protection agencies with the information they need to evaluate the resources required to cope with a major disaster, even if the scenario is very unlikely. Exercises are held periodically to familiarise all stakeholders with the good practices that should be implemented when the IOP or SRP is triggered.

In addition to the risk of a major accident, the operator must also manage the day-to-day health and safety risks. The inventory of occupational risk situations and prevention measures must be recorded in a report that can be scrutinised by the relevant authorities.

Finally, because of the hazardous nature of the products involved, the operator must carry out an impact study to measure the effects of its activity and potential accidents on the natural environment. The results of this study are also scrutinised by the relevant authorities.

The following section describes the case study site and the at-risk activity that was used to demonstrate the systemic modelling approach.

2.1.1 The Industrial Site and Its Activity

The industrial site is located in a suburban area of a French town. The surrounding area is fairly densely populated (over 6000 residents within a radius of 1 km) and it is situated close to a motorway (150 m), a railway line (20 m) and several commercial and leisure activities (e.g., shopping centres, businesses and a cinema are all within a radius of 2 km). The local population level remains almost constant throughout the year. Two schools, administrative buildings and a care facility are situated approximately 1 km from the factory.

The site is near a watercourse that a hydrological study of the area has identified as potentially subject to low-level flooding. Seismic risk is low and, as there are very few wooded areas nearby, the risk of forest fires (and associated damage) is very low.

In accordance with current regulations, the site operator has established a Safety Management System (SMS). The management have put in place a procedure to learn from past experiences. To facilitate the development of a local safety culture, the knowledge of previous accidents and incidents is made available for re-use as part of a continuous improvement process covering safety, training, information and staff awareness (Kjellen 2000).

On site, there are high-risk activities associated with the chlorine discharge bay and the line that is used to supply chlorine to the workshop where plastics are made. The company has explored the removal of chlorine from its processes, but it currently remains indispensable in the manufacture of plastics. The effects of chlorine on humans and the environment, and the study of the prevention of accidents—the accidentology—related to its use are described below.

2.1.2 Chlorine: Its Physico-Chemical, Toxicological Properties and Accidentology

Chlorine (chemical symbol, *Cl*) is a member of the halogen family, and has an atomic number of 17. It exists in gaseous form as dichlorine (Cl_2) molecules at 0 ° C and standard atmospheric pressure (1 atm or 1.01 bar). In gaseous form, chlorine is greenish-yellow and, because it is denser than air (3.2 g/l at 0 °C), it tends to form clouds close to ground level. Its smell is pungent and can be detected at a concentration below 1 ppm[1] (NIOSH 2007 and NIOSH web site). It is sparingly soluble in water (0.7 mg/100 ml at 20 °C).

Although chlorine gas is inflammable, its high reactivity with many other products (both organic and inorganic) means that it can cause explosions and fires. It reacts with water, notably in humid conditions, to form hypochlorous oxide, which can corrode metal pipes and steel tanks. High concentrations of chlorine in the atmosphere and the aquatic environment can lead to serious problems and even the destruction of animals and plants.

Chlorine is an irritant that can lead to the breakdown of cells. Its effects on human health are related to its irritant properties (see Table 2.1). At low concentrations (less than 15 ppm), it very quickly irritates the eyes, the skin and the cells of

[1]ppm = Part per million: One unit of a substance contained in a million units of another substance.

Table 2.1 Effects of chlorine on human health as a function of inhaled concentrations according to EUROCHLORE (INERIS 2004; EuroChlor 2005, 2008)

Concentration (ppm)	Effects
>1	Olfactory threshold
3–5	Upper respiratory tract irritation
5–8	Eye and upper respiratory tract irritation
15–20	Severe irritation of upper respiratory tract
30	Dyspnoea, nausea, vomiting
40–60	Pulmonary oedema
>50	Coma, death (depending on duration and concentration of exposure)

Table 2.2 Significant Lethal Effects (SLE), Initial Lethal Effects (ILE) and Irreversible Effects (IE) thresholds expressed in ppm as a function of the duration of exposure of individuals in minutes (INERIS 2008)

	Time in minutes			
Concentration (ppm)	1	10	20	30
SLE	1082	324	226	183
ILE	910	280	200	160
IE	110	41	30	25

the upper respiratory tract (nose and throat), resulting in tearing, a runny nose and a transient cough with no long-term clinical consequences. At concentrations greater than 30 ppm, it may cause dyspnoea (difficulty in breathing, suffocation), albeit with no long-term consequences if suitably treated. In higher concentrations (about 1000 ppm over 1 min), pulmonary oedema can occur in a few minutes, leading to death.

Table 2.2 shows the lethal and irreversible effects thresholds of chlorine on humans as a function of the length of exposure. At a concentration of 1000 ppm for 1 min, 1% of the exposed population has a high probability of death due to inhalation.

The chlorine accidentology (MEDD 2004) is based on an analysis of 557 accidents (including 251 in France) involving chlorine. Of these 557 accidents, 35 (four in France) were responsible for the deaths of 240 (43 in France) people. The chemical industry was involved in 10 of the 24 accidents in France with a known cause; land transport was involved in nine. The causes of these accidents were the result of several types of failure: technical (rupture of a tank or pipeline, leakage, etc.); organizational (design or process defects, poor maintenance, etc.); and human factors (poor manoeuvres, mixing incompatible substances, inappropriate response to emergency situations, etc.). Before analysing the risks associated with the use of chlorine on the site, we describe the activities associated with the discharge and transfer of chlorine.

2.1.3 Discharge and Transfer Activities

Liquid chlorine arrives at the discharge bay (Fig. 2.2) in wagons at a pressure of 6 bar. The facility holds a maximum of two full wagons at any one time. Each wagon, which contains between 55 and 59 tonnes of chlorine, is routed to one of the two bunkers for discharge. The two bunkers operate alternately because of the time needed to connect and empty the wagons. The chlorine is transferred from the wagons to the process workshop, where it is used in gaseous form, at a pressure of 2.5 bar, in the synthesis of plastics. This organization of the discharge and transfer of chlorine enables the plant to maintain a sufficient, continuous supply of chlorine for the production of plastics.

Discharge takes place in hard, sealed bunkers equipped with a ventilation system that slightly reduces the air pressure inside. The connection, emptying, and disconnection operations are carried out in an enclosed space in which chlorine leak detectors continuously analyse the air quality. If a leak equivalent to a 5 ppm concentration in the atmosphere occurs, the chlorine gas is aspirated and redirected to a deposition column (not shown in Fig. 2.2) containing sodium hydroxide, where it is neutralized.

The bunker's air temperature is maintained at 35 ° C by a heating system consisting of a steam circuit that runs from a boiler in another of the factory's workshop. This temperature increases the internal pressure of the wagon to between 8 and 9 bar. Under this pressure, liquid chlorine collects in channels and is directed to the chlorine transport line, which supplies it to the process workshop. This method allows chlorine to be extracted without the need to introduce an inert gas. It takes 3–4 days to fully discharge a wagon, depending on the consumption rate. A chlorine

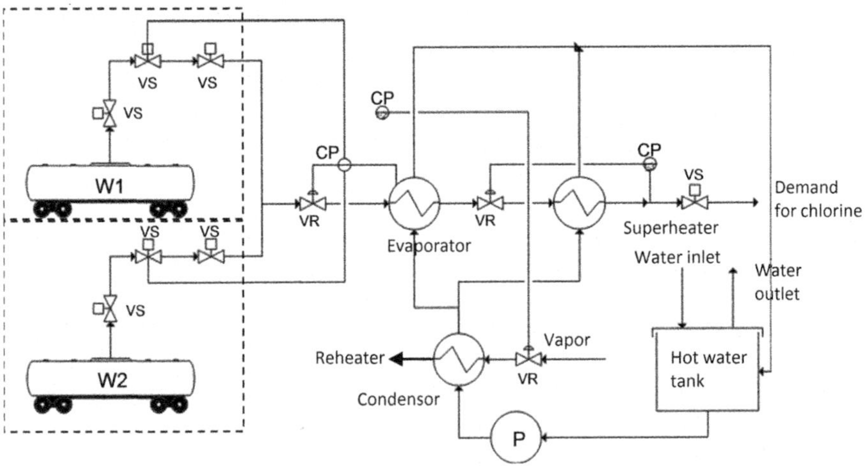

Fig. 2.2 General overview of the chlorine line. *W* wagon, *VS* safety valve, *VR* control valve, *CP* pressure sensor, *P* Pump

wagon is replaced when the line pressure indicates that the wagon is nearly empty. Switching between an empty wagon and a full one takes about 90 min.

When the wagon enters the bunker it is first stabilized, and connected to a 25 mm pipe that conducts the liquid chlorine to the evaporator. The pipes carrying the liquid chlorine are situated in the bunkers to ensure containment and to minimise their length. The section of the pipeline that is used to transfer the chlorine gas is situated outside the bunker, in line with best practice (EuroChlor 2005, 2008).

An evaporator is located in a building adjacent to the bunker, along with the previously mentioned boiler. In the evaporator, the liquid chlorine, which leaves the wagon at 35 °C, becomes gaseous and the superheater increases its temperature to 40 °C to compensate for any heat loss as it travels through the supply pipeline. This is important because any drop in temperature could cause the chlorine to condense in the pipeline. The control valve that is upstream of the evaporator ensures a constant, regular inflow of liquid chlorine.

The evaporator, like the superheater, relies on a closed circuit of hot water, which is itself heated by a steam heat exchanger. The hot water circuit is regulated by the rate of steam flow. The water temperature, which is 36 °C in normal operation, is regulated at the evaporator outlet by a steam valve. To maintain a pressure of 2.5 bar at the production workshop, a valve is placed before the superheater.

All-or-nothing safety valves that are completely open or closed when activated, are placed along the line and can be controlled by the chlorine pressure, by the triggering of leakage alarms, and by other safety sensors identified in the risk analysis. In this way, the transport and safety functions are kept separate.

2.2 Contribution of STELLA® to the Dynamic System Model

The STELLA® software package can be used to simulate the dynamic operation of a complex system, such as an industrial facility. This software is based on fundamental systemic principles (Chap. 1), which attempt to describe systems in terms of their own structures (identification and linking of sub-systems) rather than from their initial conditions. The structural description is based on the definition of constituent elements (subsystems) modelled by stocks or reservoirs. Interactions between stocks are modelled by flows and auxiliary control variables.

The STELLA® modelling environment is presented below, followed by a detailed description of the three nodes of the chlorine line that were modelled: the discharge workshop; the chlorine transfer pipeline; and the heating system. The first scenario represents normal operating conditions; this is followed by a focus on two hazardous events involving chlorine leaks.

2.2.1 STELLA®: A Systemic Modelling Environment

The advent of personal computers in the 1970s led to wider availability of software tools that could be used by the designers of systemic models, who are not expert programmers. These tools have interfaces that translate natural language into graphs composed of icons. These graphs are then used to write the equations that form the foundations for mathematical modelling and simulation calculations. Software publishers in the domain offer a range of packages; the best-known, and most widely used are STELLA® and its twin, I Think (developed by Isee Systems), Vensim (Ventana Systems Inc.), AnyLogic (AnyLogic) and Powersim Studio (Powersim Software SA). Research centres have also developed their own software platforms. These include Simile (Simulistics Ltd) originally developed by the Institute of Ecology and Resource Management at the University of Edinburgh.

STELLA®—only tool that will be discussed here—can be used to create models based on the principles of system dynamics. The environment supports three levels of model—each providing a different level of abstraction—and the simulation parameters are explored using the package's menus and tools. The three levels of models are:

- The interface: This level is designed for people who use the models (who may not be the people who created them). They can familiarise themselves with the model, run simulations and observe the results. At this level, real objects are represented in sufficient detail to allow simulation results to be explored in breadth and depth. It effectively provides a dashboard that allows the user to define the initial conditions or constraints on simulations as a function of predefined or tested scenarios.
- The diagram: This level is split into two screens (Map and Model) depending on how complete the model is. It is used to develop a diagram of the system using Forrester's (1961) principles: stocks are represented by rectangles; flows are represented by valves; sources and sinks by clouds; auxiliary variables by circles; material and energy flows by thick arrows; and information flows by fine arrows. Figure 2.3 shows the implementation of Forrester symbols in STELLA®. This level of detail allows the model's designer to develop their diagram by dragging and dropping the available objects, and establishing causal connections between them. Once these causal relationships have been established, STELLA® helps the designer to develop the mathematical functions and run the calculations that simulate the system's dynamics. The equation that calculates a value for a selected variable must contain the set of relevant variables. STELLA® helps the designer to create a simulation—a *quantified* expression—starting from a *qualitative* structure of the model (the Forrester diagram).
- The equations: This level displays the list of equations for each of the model's variables, including: state variables; flow variables; and auxiliary variables—either as tabulated functions or unique constants. This level can also display the model designer's comments that describe the equations in natural language, and

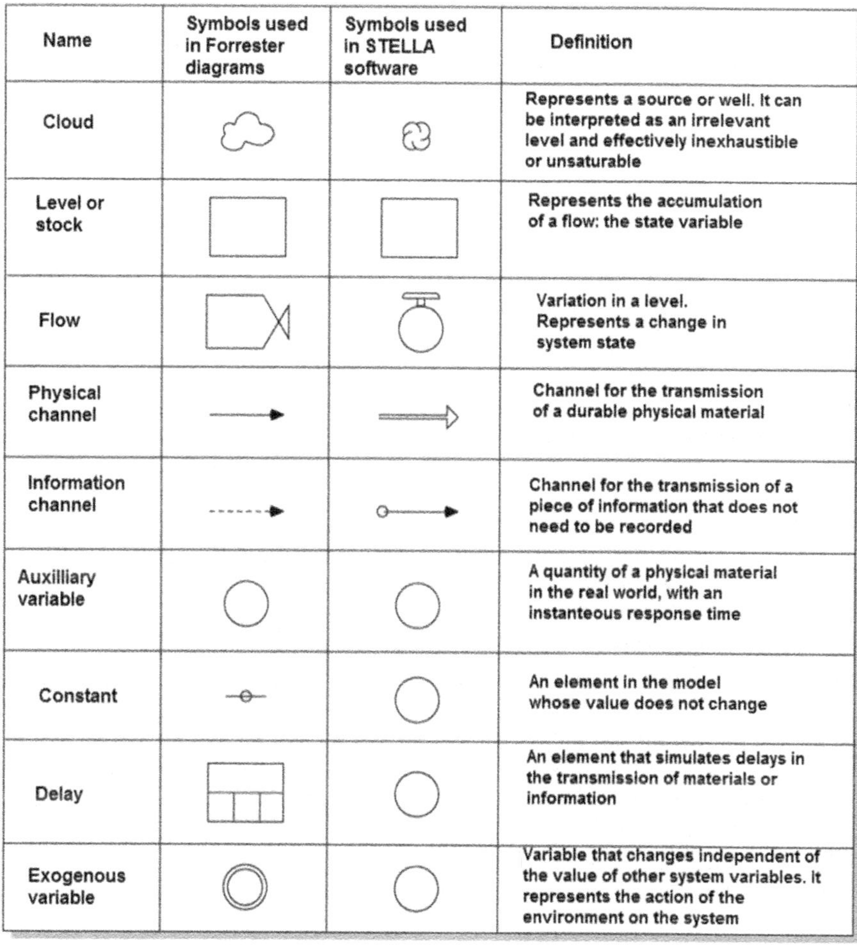

Fig. 2.3 The Forrester graphical language and its implementation in STELLA®. (Translated from Chery J.-P. and Smektala G., 2004, p. 534, following Aracil J., 1984)

provides users with valuable information on the assumptions, choices and sources of information that were used to develop the equations.

Tools like STELLA® make it easier for non-programmers to develop computer models. STELLA® effectively provides a software engineering environment that can be used to produce a particular program (the model). The interface is composed of icons that are presented as generic elements that make up a causal graph. It is designed to resemble a blank canvas for the development of draft models based on Forrester principles (Fig. 2.4).

Although STELLA® has evolved over time, and new features have been introduced, the central principle—of creating the causal graph as the basis for modelling—remains unchanged.

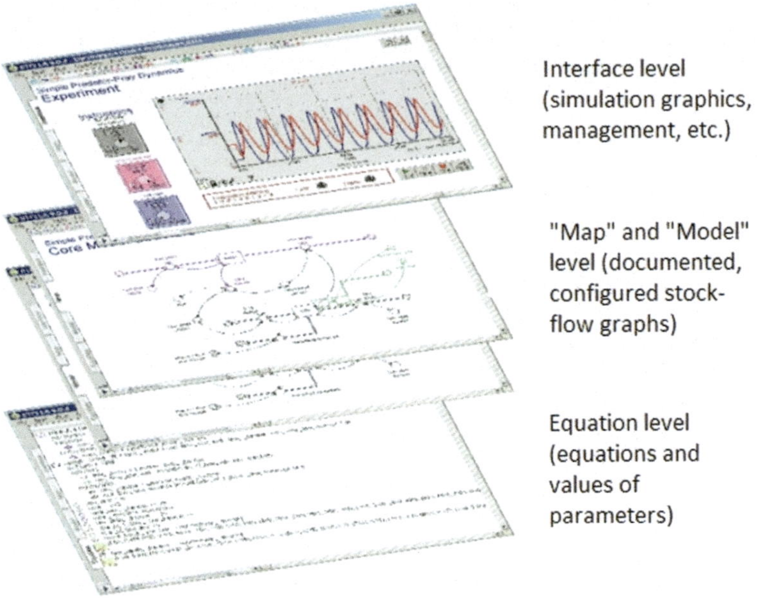

Fig. 2.4 The various levels of design and representation of a model in STELLA®

Fig. 2.5 Simple causal graph of two regulatory feedback loops

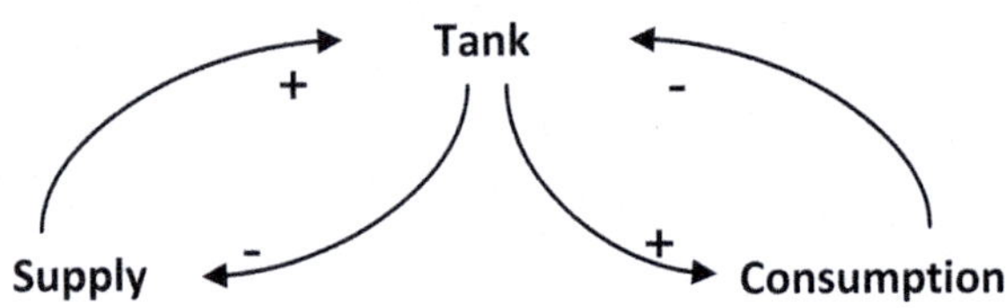

The design of a model begins by constructing a stock–flow diagram starting from the causal graph schema provided by the software. A simple illustrative example is described below.

The modelled system is a tank containing a product (chlorine), which is supplied by a feed (supply) action, and emptied by a consumption action. The tank level at time t depends on its level at time $t - 1$ and the net difference between feed and consumption. Information about the level at each instant is used to change the feed (up or down) and consumption (by varying it depending on the quantity remaining in the tank). This can be represented by a simple causal graph (see Fig. 2.5).

This causal graph can then be interpreted in the model interface of STELLA® by choosing a stock rectangle to represent the tank, and two flow variables to represent feed and consumption respectively (see Fig. 2.6).

Each of the question marks ("?") in Fig. 2.6 indicates either that a particular variable is undefined, or that there are no explicit equations that can be used to run simulations. The stock–flow diagram is completed by connecting the information

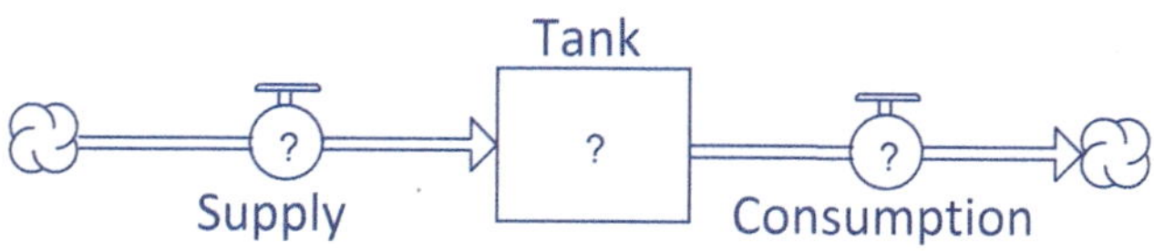

Fig. 2.6 First stage of the construction of the stock–flow diagram

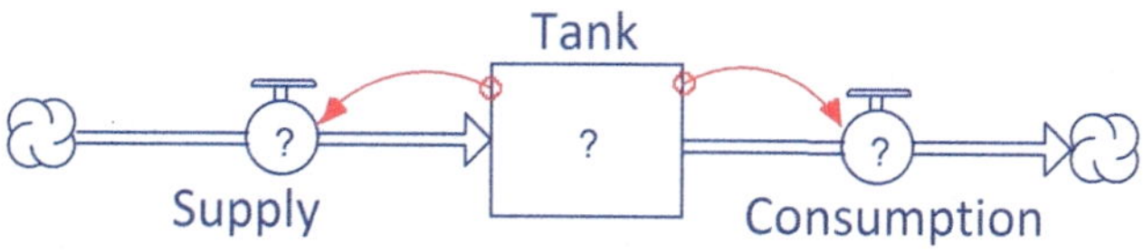

Fig. 2.7 Second stage of the construction of the stock–flow diagram

channels between the tank and each of its streams so that their values are regulated by the level of product in the tank (Fig. 2.7).

Each variable is then linked to those that it has a cause–effect relationship with, and the STELLA® interface displays a configuration panel. For the stock variable *tank*, the user can set the initial value, e.g., to the amount of chlorine in the tank, e.g., 50 l.

Each flow variable can also be defined via a dialog box (Fig. 2.8). These definitions can be conditional, e.g., the tank supply value is only set (to the quantity of chlorine that will be added to the tank in the next step) when the tank level falls below some threshold level and hence triggers an inflow of chlorine.

The function can be written by using one of STELLA®'s built-in functions, in the form:

IF *tank* < 10 THEN 40 ELSE 0

If the level of the tank falls below 10 l, 40 l are added. This implicitly means that when the level is higher than 10, the input variable, *supply* takes a value of zero. In each time step, the output variable *consumption*, is assumed constant at 2 l, in normal use. Given an initial value of 50 l, the tank will empty in 25 time steps, in the absence of any other supply. If the level falls below 10 l, consumption decreases to 1 l per time unit. It is therefore a conditional equation that takes the form:

IF *tank* < 10 THEN 1 ELSE 2

A temporal graphical representation of the system (Fig. 2.9) shows the variation in the level of the tank and flow variables as part of the simulation interface.

This is a simplified overview of how to build and run simulations in STELLA®. More detailed descriptions can be found elsewhere (e.g., see Deaton and Winebrake's (2000) bibliography). The various sub-system (or node) models associated with the discharge and transportation of chlorine are presented next, focusing on the definition of variables and their relationships.

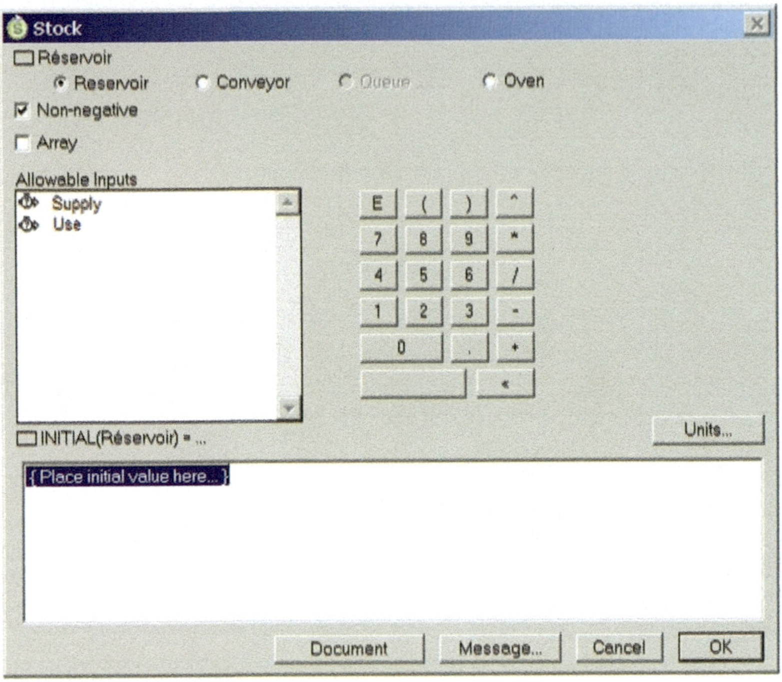

Fig. 2.8 Interface to set the initial value of the stock variable

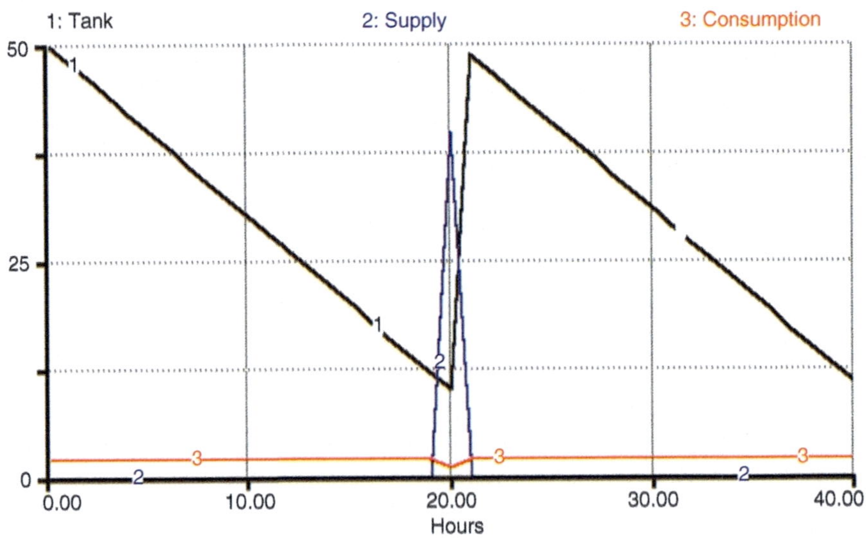

Fig. 2.9 Graphic of the result of a simulation in STELLA®

2.2.2 *General Causal Graph of the Discharge and Transfer System*

By following the main stages of modelling, the causal graph is developed and refined from an abstract level to the final, most detailed level.

2.2.2.1 Causal Graph at the Aggregate Level

The case study focuses on the three main elements involved in the delivery of chlorine, from offsite, to its subsequent reconditioning for use in the production of plastics. These three elements, organised as shown previously, in Fig. 2.2, are: the discharge bay; the chlorine line; and the heating component. Together they regulate the supply of chlorine to the production unit.

The relationships between the three elements can be established using information about the organization of the industrial site:

- The discharge bay provides chlorine to the line at a rate regulated by its internal organization—two wagons are alternately replaced and emptied—and the associated safety measures.
- The chlorine line provides information to the discharge bay about downstream production needs, which involves maintaining a relatively constant flow of chlorine.
- The heating component controls the physical conditions of this exchange. It comprises a heat exchanger that regulates the temperatures of water and chlorine in order to maintain the chlorine in a stable gaseous state, a balance that is maintained by the evaporator and the superheater. In turn, the heat exchanger receives information about the state of chlorine in the chlorine line.

From this we can establish an aggregated causal graph at the most abstract level (Fig. 2.10). Each arrow indicates the nature of the causal relationship: the discharge bay provides chlorine to the chlorine line, it therefore tends to increase the presence of chlorine in this component (+), but delivery is regulated by the need for a regular flow in the chlorine line, which dictates the need to limit delivery to a defined level (−).

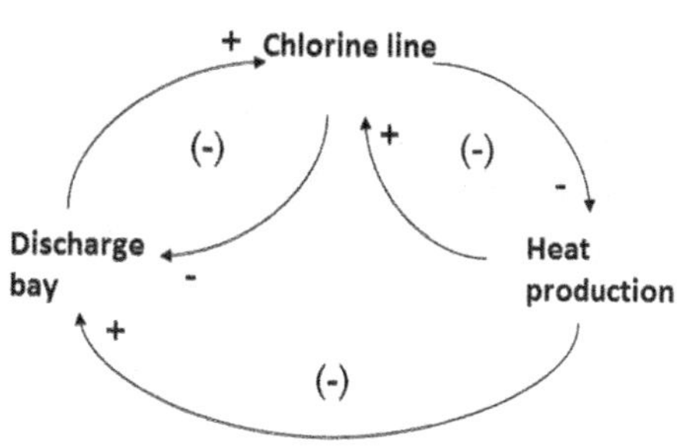

Fig. 2.10 General causal graph at the aggregate level of the discharge bay and the transfer of chlorine at the industrial site

The chlorine line interacts with the superheater in the same way: the superheater tends to increase the flow of chlorine in the line (+), and the line responds with information that regulates heat output to a level that ensures a regular flow of chlorine (−). A control unit receives information from the chlorine line and the superheater, which it uses to safely manage the emptying of wagons at an appropriate rate.

There is therefore a causal link between heat production and the discharge bay, shown by a (+). This link is asymmetrical: the heat generator should only ensure the gaseous state and rate of chlorine flow in the line. This means there is no need to include a causal arrow from the discharge bay to the heat generator.

At this very general level, the system consists of three regulatory loops that ensure the overall safe delivery of chlorine for plastics production. This simple model is an aggregate view of the system, and forms the basis for the full model, which is constructed from quantifiable variables that can be linked using these feedback loops. This is an iterative process that involves decomposing each of the major components into finer-grained elements, and maintaining feedback loops by linking the new elements within each segment of the model and between the three aggregated sectors.

2.2.2.2 Disaggregated Causal Graph

Figure 2.11 shows the variables used in the model and their causal relationships. It does not show configuration parameters such as constants, although they can be shown in stock–flow diagrams: they are not considered causal elements in feedback loops, but are used in solving equations when simulations are run.

The causal graph provides a qualitative perspective on the relationship between the elements of the modelled system, but also retains those elements that can be expressed numerically.

This type of representation allows feedback loops to be developed in detail. For example, there is a fairly simple feedback loop in the chlorine discharge bay: the element *chlorine wagon 1* (one of the two available wagons) is affected by its availability to be connected to the chlorine line and, therefore, to supply it. Its availability is the positive cause of the physical presence of the wagon. When the wagon is connected to the line, however, it is no longer in an available state (which is when wagon 2 becomes available). This is therefore a negative control feedback loop (−), as shown in Fig. 2.12. The discharge bay includes another feedback loop. The quantity of chlorine to be delivered is dependent on there being a wagon available, which can be connected (for example, *chlorine wagon 2* in Fig. 2.12). However, this requires a permutation order that validates the quantity to be delivered, and consequently sets the quantity required to 0. This is also a negative feedback loop (−), this time between four elements, two of which are involved in the negative loop for wagon 1, described above.

Once the detailed causal graph has been established, the description of the elements and their links can be translated into a stock–flow diagram, using STELLA®. Forrester symbols and numeric expressions (such as differential equations, functions

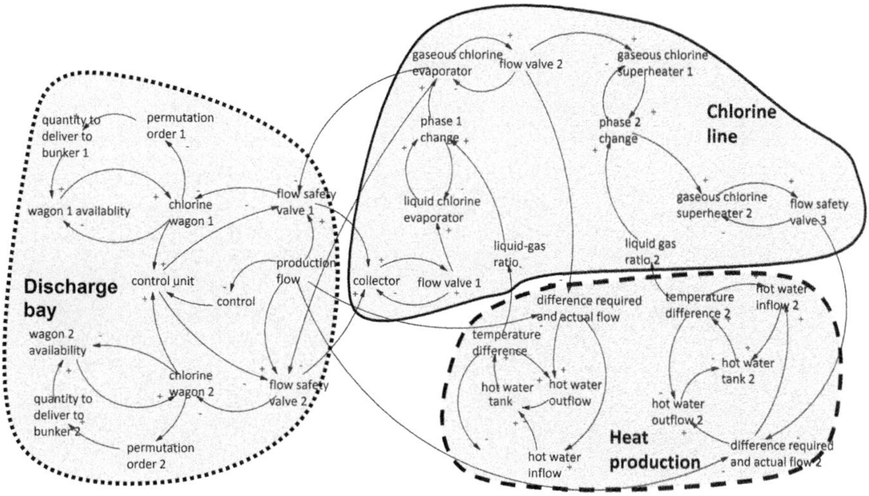

Fig. 2.11 Detailed causal graph at the disaggregated level showing the discharge and transfer system at the industrial site

and parameters) are used to describe the available choices and to support the choice of variables. The following sections expand on these descriptions starting with the three principal areas of the site: the discharge bay; the chlorine line; and heat production.

2.2.3 Modelling the Discharge Bay

The two chlorine discharge bays were modelled using the following elements:

- Flow variables (*wagon_1_availability* and *wagon_2_availability*) represent the availability of the chlorine wagons for emptying. When wagon 2 is being emptied (i.e., when it contains more than 1000 kg of chlorine), the permutation order for wagon 1 is not activated (its value is 0). When wagon 2 is empty, the permutation order for wagon 2 is activated (its value is set to 1) and wagon 1 is connected to the pipeline for emptying.
- Stock variables (*chlorine_ wagon_1* and *chlorine_ wagon_2*) represent the quantity of chlorine in each (full) wagon at time *t*. The value of this stock variable varies from 55 to 59 tonnes.

Definition of the variable *wagon_1_availability*
IF (*chlorine_ wagon_1* < *permutation_threshold_of_the_wagon*) THEN *quantity_to_deliver_to_bunker_1*) ELSE (0))

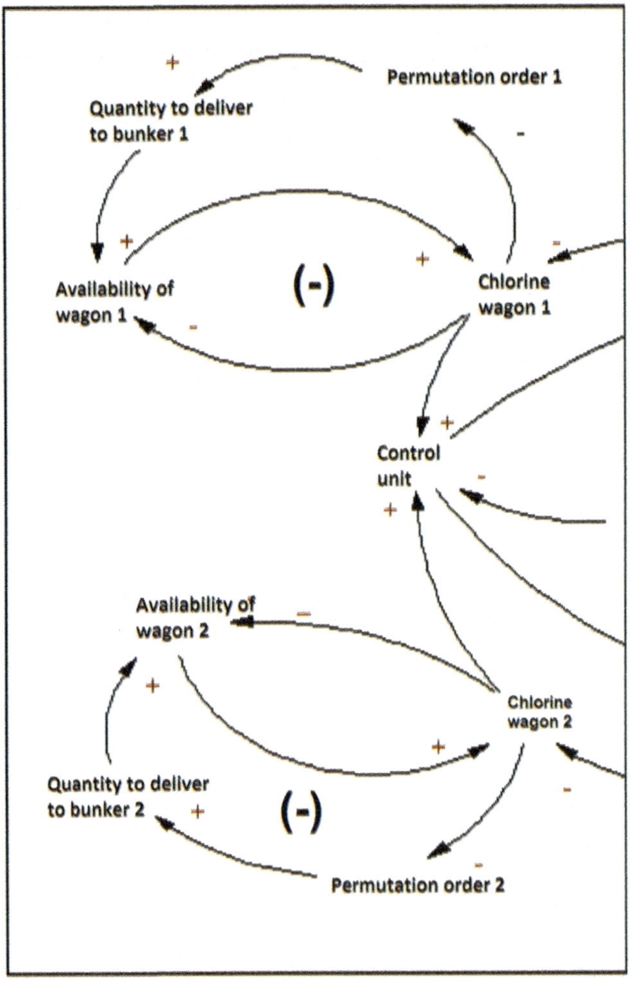

Fig. 2.12 Two feedback loops in the detailed causal graph of the discharge system

- The flow variable, *flow_safety_valve_1*, placed after the output of the stock variable *chlorine_ wagon_1* represents the flow of chlorine to another stock (*collector*). Valve 1 is opened when the control unit orders the discharge of chlorine from wagon 1; this valve is closed when wagon 2 is being discharged.

Definition of the variable *flow_safety_valve_1*
IF (*control_unit* = *1*) THEN (IF (*gaseous_chlorine_evaporator* > 2000) THEN (0) ELSE (*production_flow*)) ELSE (0)

- The stock variable (*collector*) represents the quantity of chlorine needed for production. This stock corresponds to the quantity of chlorine passing through the pipe at the outlet of either valve 1 or valve 2, depending on which wagon is being discharged. It represents the entry point for chlorine in the *Chlorine line* node. The quantity that is received varies depending on demand, i.e., the quantity needed for production.
- Auxiliary variables simulate external actions and represent the following parameters:
  - The permutation order (*permutation_order_1* and *permutation_order_2*) of the wagons depends on the quantity of chlorine they contain. When wagon 1 is being discharged (i.e., the quantity of chlorine in the wagon is 1000 kg or more), the permutation order is set to 0; when the quantity of chlorine in wagon 1 is less than 1000 kg, its order changes to 1.

Definition of the variable *permutation_order_1*
IF (*chlorine_ wagon_1* < *permutation_threshold_of_the_wagon*) THEN 1 ELSE 0

  - The permutation threshold (*permutation_threshold_of_the_wagon*) represents the level of chlorine in a wagon that triggers a change to the other wagon. This threshold is set at 1000 kg.
  - The quantity of chlorine variables (*quantity_to_deliver_to_bunker_1* and *quantity_to_deliver_to_bunker_2*) show how much chlorine should be delivered to each bunker. This quantity assumes that there is a delay related to the time taken to (empty and) swap wagons, i.e., about 48 h ±5 h.

Definition of the variable *quantity_to_deliver_to_bunker_1*
DELAY ((*permutation_order_1* x *wagon_quantity*), (*charging_time*)

  - The time taken to swap wagons (*charging_time*) is about 48 h ± 5 h. It corresponds to the time need to remove an empty wagon and receive a full wagon. The *NORMAL* function is used to generate random numbers from a normal distribution with a specified mean (48 h) and standard deviation (5 h).

Definition of the variable *charging_time*
NORMAL (48, 5)

  - The quantity of liquid chlorine (*wagon_quantity*) supplied in each wagon varies between 55 and 59 tonnes. The *RANDOM* function is used to simulate the variability in this quantity for each wagon.

Definition of the variable *wagon_quantity*
RANDOM (55000, 59000, 1)

- Production flow (*production_flow*) represents the rate of chlorine flow at the point where it exits the *Chlorine line* node. The rate required by the production unit is set at 770 kg/h.
- The *control* variable represents the quantity of chlorine that should be delivered per hour by the pipeline in normal operating conditions. A value of 770 is set as the flow threshold. Values below this threshold are considered abnormal and indicate an event such as a leak.

Definition of the variable *control*
IF (*production_flow* < 770) THEN 1 ELSE 0

- The control unit (*control_unit*) is used to select which wagon to discharge by activating the appropriate safety valves (safety valve 1 or 2). Flow safety valve 1 is opened when the quantity of chlorine in wagon 1 exceeds 1000 kg and the quantity of chlorine in wagon 2 has fallen below 1000 kg. Similarly, safety valve 2 is opened when wagon 2 contains more than 1000 kg of chlorine, and the quantity in wagon 1 has fallen below 1000 kg. Safety valves 1 and 2 are closed when the quantity of chlorine in both wagons falls below 1000 kg.
- If the quantity of chlorine in wagon 1 is less than wagon 2, safety valve 1 remains open, and wagon 1 is completely emptied. Conversely, if wagon 2 holds less chlorine than wagon 1, the wagon 2 valve remains open and wagon 2 is completely emptied. If both wagons contain the same quantity of chlorine, wagon 2 is emptied first. In all cases, only one valve can be actuated at any one time.

Flow and stock variables and auxiliary variables are linked together by connectors that control the system, simulating feedback loops. Each wagon is modelled by the quantity of chlorine it contains: initially 55–59 tonnes. At any particular moment only one wagon can be empty. When one wagon is empty, a switchover takes place so that the other wagon can be emptied.

Definition of the variable *control_unit*
IF (*control* = 1) THEN 0 ELSE
IF ((*chlorine_wagon_1* < *permutation_threshold_ of_the_wagon*) AND (*chlorine_wagon_2* > *permutation_threshold_of_the_wagon*)) THEN 2
ELSEIF ((*chlorine_wagon_1* > *permutation_ threshold_of_the_wagon*) AND (*chlorine_wagon_2* < *permutation_threshold_of_the_wagon*)) THEN 1
ELSEIF ((*chlorine_wagon_1* < *permutation_threshold_of_the_wagon*) AND (*chlorine_wagon_2* < *permutation_threshold_of_the_wagon*)) THEN 0
ELSEIF (*chlorine_ wagon_1* < *chlorine_wagon_2*) THEN 1
ELSE 2

The model of chlorine discharge operations (See Fig. 2.13) uses the variables listed in Table 2.3. The table includes details of the interrelationships between variables as presented in STELLA®.

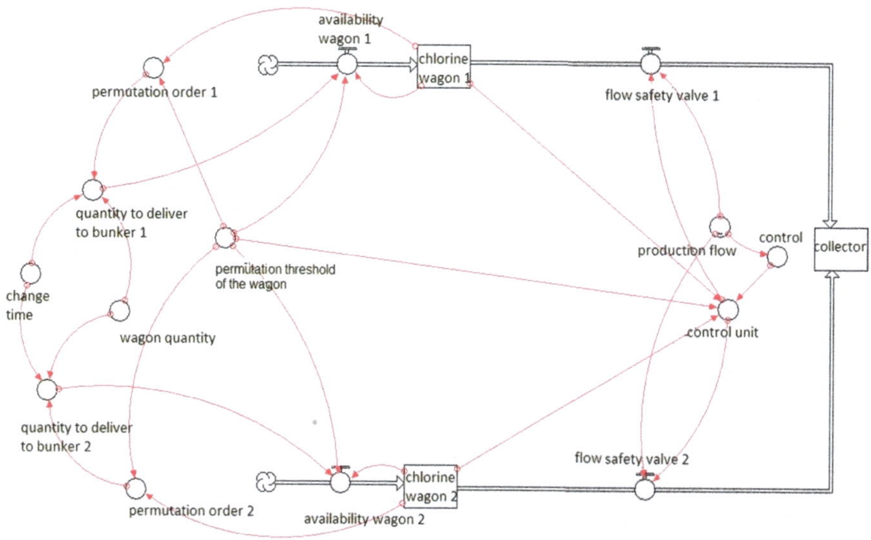

Fig. 2.13 Model of the chlorine discharge system

Table 2.3 Definition and function of variables used to model the chlorine discharge unit

Chlorine discharge unit		
Variable name	Definition	Function
wagon_1 availability	IF (*chlorine_tank_1 < permutation_threshold_of_the_wagon*) THEN (*quantity_to_deliver_bunker_1*) ELSE 0	Chlorine available in transfer tanks
chlorine_ wagon_1	55,000 or 59,000	Capacity of the transfer tank
flow_safety_ valve_1	IF (*control_cabin* = 1) THEN IF (*gaseous_chlorine_ evaporator* > 2000) THEN 0 ELSE (*production_flow*) ELSE 0	Simulates the chlorine flow toward the collector
collector	Stock	Quantity of chlorine leaving safety valves 1 and 2
permutation_ order_1	IF (*chlorine_wagon_1 < permutation_ threshold_of_the_ wagon*) THEN 1 ELSE 0	Depends on the quantity of chlorine remaining in each tank
permutation_ threshold_of_ the _wagon	1000	Chlorine threshold that activates the change of tank

(continued)

Table 2.3 (continued)

Chlorine discharge unit		
Variable name	Definition	Function
quantity_to_deliver_to_bunker_1	DELAY ((*permutation_order_1* * *wagon_quantity*), *charging_time*)	Quantity of chlorine to be delivered to bunker 1
charging_time	NORMAL (48.5)	Time to change a tank (48 ± 5 h)
wagon_quantity	RANDOM (55,000, 59,000, 1)	Quantity of liquid chlorine delivered by the rail network (55–59 tons)
production_flow	770	Chlorine demand equal to 770 kg/h
Control	IF (*production_flow* > 700), THEN 1 ELSE 0	Controls chlorine flow rate
control_unit	IF (*control* = 1) THEN 0 ELSE IF ((*chlorine_wagon_1* < *permutation_threshold_of_the_wagon*) AND (*chlorine_wagon_2* > *permutation_threshold_of_the_wagon*)) THEN 2 ELSEIF ((*chlorine_wagon_1* > *permutation_threshold_of_the_wagon*) AND (*chlorine_wagon_2* < *permutation_threshold_of_the_wagon*)) THEN 1 ELSEIF ((*chlorine_wagon_1* < *permutation_threshold_of_the_wagon*) AND (*chlorine_wagon_2* < *permutation_threshold_of_the_wagon*)) THEN 0 ELSEIF (*chlorine_wagon_1* < *chlorine_wagon_2*) THEN 1 ELSE 2	Selects the tank to be emptied by the successive activation of the safety valves

2.2.4 *Modelling the Chlorine Line*

The chlorine line that turns liquid chlorine into gas and routes it to the production unit is also modelled by stock, flow and auxiliary variables. These variables are presented in the order they appear in the model shown in Fig. 2.15.

- *flow_valve_1*

Flow variable showing the quantity of liquid chlorine passing from the collector to the evaporator.

- *liquid_chlorine_evaporator*

Stock variable showing the quantity of liquid chlorine entering the evaporator where it is converted into chlorine gas. The evaporator contains more chlorine than the quantity contained in the pipeline due to the accumulation of chlorine over time.

- *phase_1_change*

Flow variable representing the chlorine state change within the evaporator. The quantity of liquid chlorine that is fed into the evaporator is multiplied by a coefficient that simulates the transformation of liquid chlorine into chlorine gas.

Definition of the variable *phase_1_change*
liquid_chlorine_evaporator x *liquid_gas_ratio*

- *gaseous_chlorine_ evaporator*

Stock variable showing the quantity of chlorine gas that has been produced, and which will be transferred into the pipeline at the evaporator's outlet. This information is transferred back to safety valves 1 and 2 located in the discharge bunkers so that they can be opened or closed, as necessary.

- *flow_valve_ 2*

Flow variable showing the quantity of chlorine gas transferred to the superheater. The quantity of gas that is produced is used by the auxiliary variable *difference_required_and_actual_flow* in calculating the difference between production requirements and the quantity discharged.

- *difference_required_and_actual_flow*

This auxiliary variable represents the difference between the quantity of liquid chlorine that is discharged and the quantity of gaseous chlorine required for production. This data makes it possible to regulate the flow and identify possible failures.

Definition of the variable *difference_required_and_actual_flow*
production_flow – *flow_valve_2*

- *gaseous_chlorine_superheater_1*

Stock variable representing the quantity of chlorine gas in the superheater.

- *phase _2_change*

Flow variable representing the ratio between the quantity of chlorine gas and the residual liquid chlorine that needs to be evaporated.

Definition of the variable *phase_2_change*
gaseous_chlorine_superheater_1 x *liquid_gas_ratio_2*

- *gaseous_chlorine_superheater_2*

Stock variable representing the quantity of chlorine gas that leaves the superheater and is transported to the production unit.

- *flow_safety_valve_3*

Flow variable showing the quantity of chlorine gas supplied to the production unit taking into account the (small) effect of potential heat losses due to the condensation of chlorine.

Definition of the variable *flow_safety_valve_3*
gaseous_chlorine_ superheater_2 x *heat_dissipation*

- *heat_dissipation*

Auxiliary variable simulating the effects of heat dissipation from the chlorine gas in the pipeline. The external pipelines are enclosed in an insulating sleeve, so losses are zero. If the insulation system becomes damaged, however, there may be minor losses (1–2%): this variable is therefore modelled by a constant function.

Definition of the variable *heat_dissipation*
1

- *liquid_gas_ratio*

Auxiliary variable simulating the relationship between the liquid and gas states of chlorine in the pipeline (Fig. 2.14). This ratio varies as a function of: the temperature of the hot water that passes through the evaporator and the superheater; and the flow rate of hot water.

Table 2.4 shows the variables used to model the chlorine line.

Figure 2.15 shows the variables used to model the chlorine line and the relationships between them as presented by STELLA®.

2.2.5 *Modelling Heat Production*

For normal operations, the production unit needs to be supplied with sufficient levels of chlorine gas. These levels are controlled through the production, regulation and circulation of hot water, which is used to convert the liquid chlorine into gas. The two key parameters needed to model the water circuit are: the hot water's flow rate, which is proportional to the chlorine flow in the evaporator; and the quantity of water in the tank, which has to be high enough to supply both the evaporator *and* the

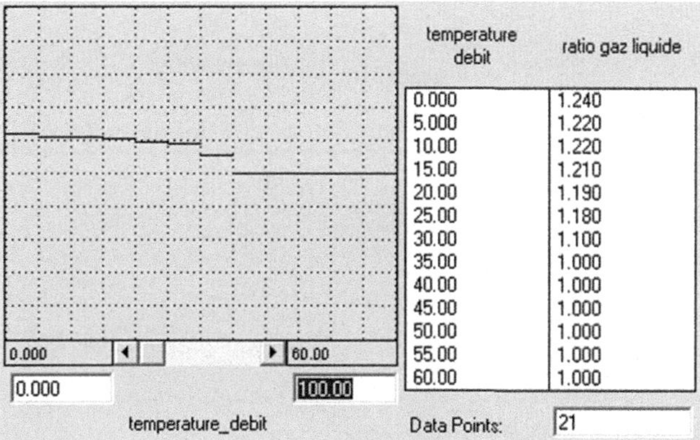

Fig. 2.14 Simulation of the ratio between the liquid and gas fractions of chlorine in the pipeline as a function of temperature and hot water flow rate in the evaporator and superheater

Table 2.4 List of variables used to model the chlorine line

Chlorine line		
Variable name	Definition	Function
flow_valve_1	Flow	Quantity of liquid chlorine passing from the collector to the evaporator
liquid_chlorine_evaporator	Stock	Quantity of liquid chlorine that enters the evaporator to effect its state change
phase_1_change	*liquid_chlorine_evaporator* × *liquid_gas_ratio*	Simulates the state change of chlorine in the evaporator
gaseous_chlorine_evaporator	Stock	Quantity of chlorine gas leaving the evaporator
flow_valve_2	Flow	Quantity of chlorine gas in transit to the superheater
difference_required_and_actual_flow	*production_flow* – *flow_valve_ 2*	Difference between the quantity of liquid chlorine discharged and chlorine gas requirements
gaseous_chlorine_superheater_1	Stock	Quantity of chlorine gas in the superheater
phase_2_change	*gaseous_chlorine_superheater_1_* × *liquid_gas_ratio_2*	Ratio between chlorine gas and residual liquid chlorine that will evaporate
gaseous_chlorine_superheater_2	Stock	Quantity of chlorine gas leaving the superheater
flow_safety_valve_3	*gaseous_chlorine_superheater_2* × *heat_dissipation*	Quantity of chlorine gas leaving the pipeline weighted by heat dissipation
heat_dissipation	1	Heat dissipation is null in this case
liquid_gas_ratio	Graph	Simulates the ratio of liquid and gaseous state chlorine in the chlorine pipeline

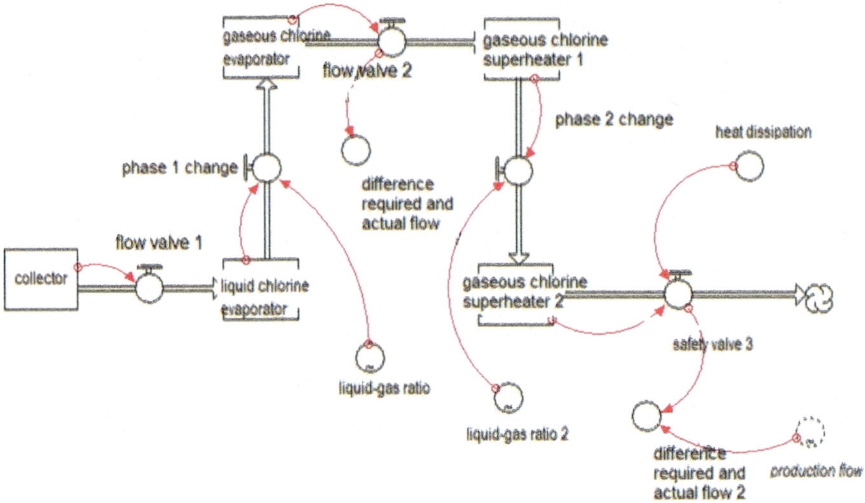

Fig. 2.15 Model of the chlorine line

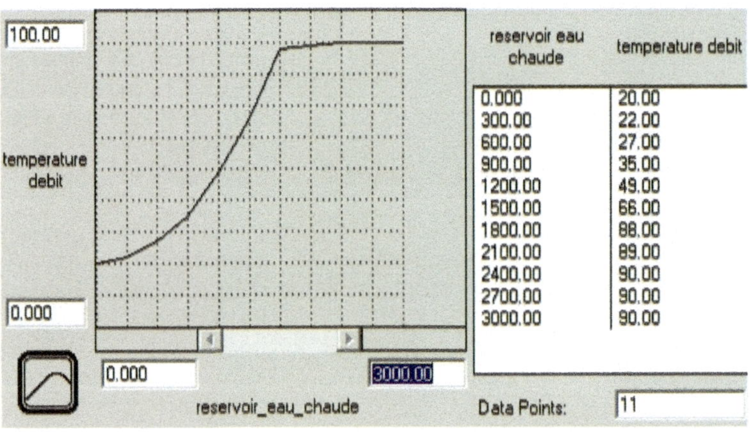

reservoir eau chaude	temperature debit
0.000	20.00
300.00	22.00
600.00	27.00
900.00	35.00
1200.00	49.00
1500.00	66.00
1800.00	88.00
2100.00	89.00
2400.00	90.00
2700.00	90.00
3000.00	90.00

Fig. 2.16 Change in the quantity of hot water in the tank as a function of flow and hot water requirements needed to effect the chlorine state change in the line associated with it

superheater. The circulation of hot water was modelled using the following parameters:

- *ratio_temperature*

This variable represents the relationship between the hot water temperature, which must be high enough to evaporate the chlorine, and the flow of hot water, which must be correctly adjusted to allow the change of state of the chlorine. Figure 2.16 shows the variation in *ratio_temperature* as a function of the quantity of hot water in the tank. The *ratio_temperature* value is passed to the auxiliary variable *liquid_gas_ratio* that controls the demand for hot water in order to ensure the change of state in the chlorine in the relevant line.

- *hot_water_inflow*

This flow variable (*hot_water_inflow_1* and *hot_water_inflow_2*) represents the quantity of hot water entering the hot water tanks for the evaporator and the superheater in a particular time interval. It is linked to the auxiliary variables *ratio_temperature* and *difference_required_and_actual_flow* to regulate the quantity of hot water needed for the change of state of the chlorine in the evaporator and the superheater.

Definition of the variable *hot_water_inflow*
IF (*ratio_temperature* < 36 AND *difference_required_and_actual_flow* > 0)
THEN 50 ELSE 0

If the temperature and flow rate are too low for the evaporator to maintain a temperature of 36 ° C, and the production levels of chlorine gas are too low, the flow and the temperature of the hot water must be increased in order to keep the production unit supplied with chlorine gas. These increases are achieved by raising the level of hot water in the tank.

- *hot_water_tank*

This stock variable (*hot_water_tank_1* and *hot_water_tank_2)* represents the amount of hot water that is available to supply the evaporator and the superheater.

- *hot_water_outflow*

This flow variable (*hot_water_outflow_1* and *hot_water_outflow_2*) represents the quantity of hot water entering the evaporator and the superheater. Like the variable *hot_water_inflow_1* it is controlled by the auxiliary variables *ratio_temperature* and *difference_required_and_actual_flow*.

Definition of the variable *hot_water_outflow*
IF (*ratio_temperature* < 47 AND *difference_required_and_actual_flow* > 0)
THEN 60 ELSE 0

Table 2.5 lists the variables that were defined to model production and heat regulation in the discharge unit and for the transfer of chlorine.

Figure 2.17 shows the general pattern of the production of water and the regulation of its temperature for the evaporator and the superheater.

Table 2.5 List of variables used to model production and heat regulation

Heat production		
Variable name	Definition	Function
ratio_ temperature	Graph	Simulates the relation between the temperature of the water, which must be high enough to evaporate the chlorine, and the flow of hot water that must be correctly adjusted to allow the state change.
hot_water_ inflow	IF (*ratio_temperature* < 36 AND *difference_required_and_ actual_flow* > 0) THEN 50 ELSE 0	Simulates the quantity of hot water entering the hot water tank.
hot_water_ tank	Stock	Quantity of hot water available to ensure the supply of hot water from the evaporator and superheater.
hot_water_ outflow	IF (*ratio_temperature* > 47 AND *difference_required_and_ actual_flow_* > 0) THEN 60 ELSE 0	Quantity of hot water entering the evaporator and the superheater.

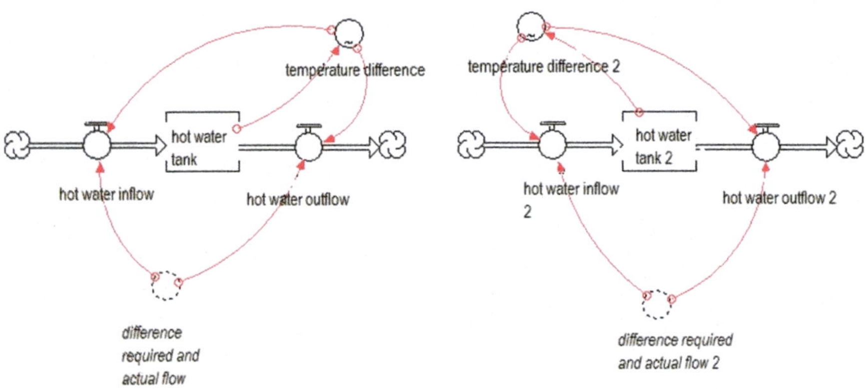

Fig. 2.17 Modelling production and the chlorine line hot water cycle: two cycles were modelled as hot water must be provided to both the evaporator and the superheater

2.2.6 *The General Model of Normal Operations*

The three nodes described above can be combined to provide a global model of the discharge of chlorine and its transfer to the production unit. The simulation of the operation of this system in STELLA® uses graphs that show the state of variables over the simulated time period which, in this case, was arbitrarily set to 400 h (just under 17 days).

There are two main types of variations that naturally occur in normal operating conditions (see Fig. 2.18). The first is the quantity of chlorine that is discharged,

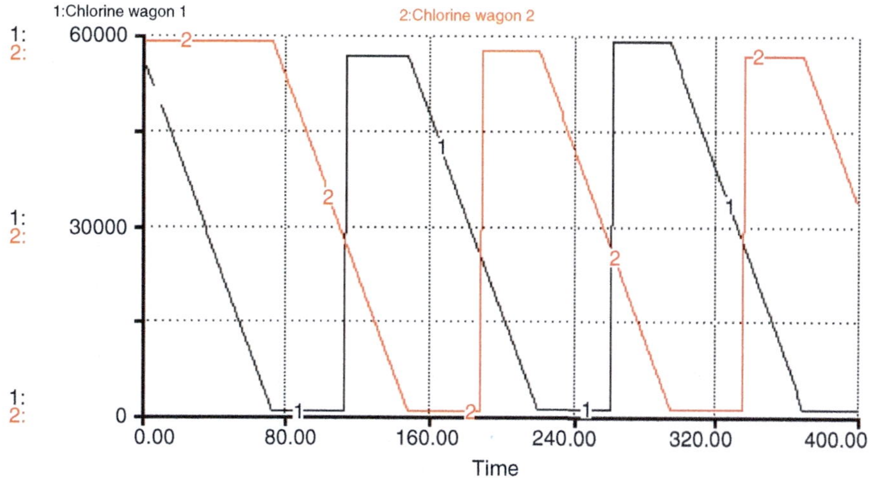

Fig. 2.18 Simulation of the discharge of chlorine wagons. 1 = chlorine wagon 1; 2 = chlorine wagon 2; the x-axis shows time in hours, from 0 to 400; the y-axis shows the quantity of chlorine in both wagons (0–60,000 kg)

which can vary between 55 and 59 tonnes. This explains the difference in levels between the two wagons when they arrive on the site.

The second change is the time it takes to discharge a wagon. This varies as a function of the quantity of chlorine delivered and production requirements. In Fig. 2.18, for example, it takes 70 h to discharge wagon 1, and about 90 h to discharge wagon 2.

Figure 2.19 shows the status of variables *permutation_order_1* and *permutation_order_2* over time. *permutation_order_1* represents the status of wagon 1. When it is being discharged, which takes 70 h, and there is more than 1000 kg of chlorine in the wagon, it is not set (i.e., has the value 0). When the chlorine level falls below 1000 kg it is activated (set to 1). *permutation_order_1* remains activated until a new wagon is connected in bunker 1. When the discharge of wagon 2 ends (at time $t = 150$ h), *permutation_order_2* is activated (set to 1) until another chlorine wagon arrives and is connected in bunker 2.

Figure 2.20 presents the flow of liquid chlorine that passes through safety valves 1 and 2 at the outlet of wagons 1 and 2. When *safety_valve_1* is open it allows wagon 1 to supply the production unit. After 70 h, this valve closes and *safety_valve_2* opens to allow wagon 2 to be discharged. The action of the two valves and the chlorine flows that depend on them follow the production unit's demands over time. In the simulation, chlorine demand is shown as 770 kg/h.

The flow of liquid chlorine passes through the stock variable *collector* before being converted into chlorine gas in the evaporator. Figure 2.21 shows the correspondence between the production unit's chlorine requirements and the chlorine flow through the two safety valves (at the inlet and the outlet of the evaporator) and *safety_valve_3*, which is at the outlet of the superheater.

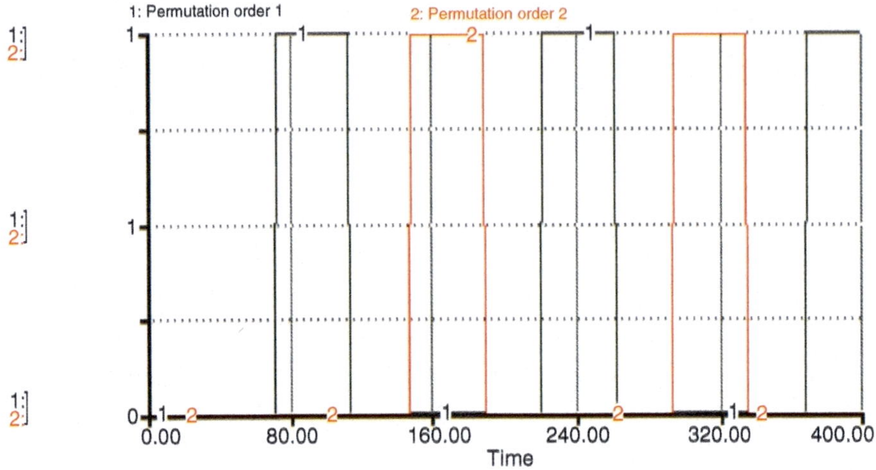

Fig. 2.19 Simulation of permutation orders 1 and 2 of wagons during discharge. The x-axis shows time, from 0 to 400 h. The y-axis is a binary variable indicating the permutation (1) or not (0) of the wagons

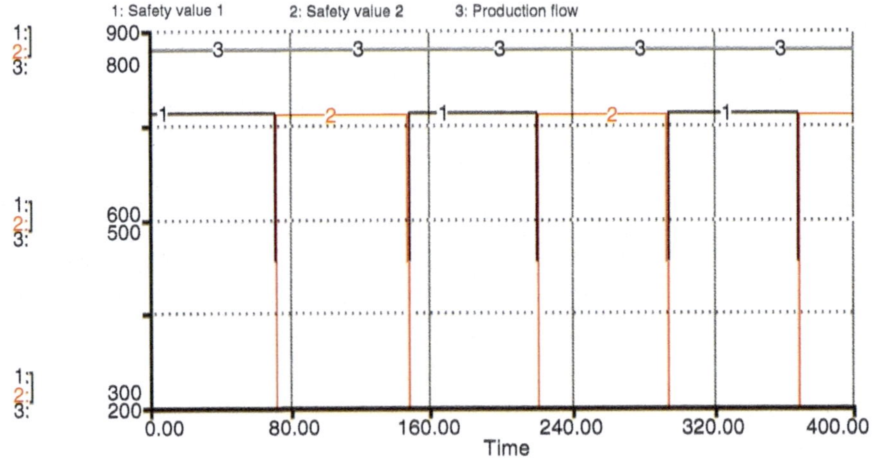

Fig. 2.20 Simulation of liquid chlorine transiting safety valves 1 and 2, and chlorine requirements of the production unit. The x-axis shows time in hours; for safety valves 1 and 2 the y-axis ranges from 300–800 kg/h of liquid chlorine; for ease of reading production requirements range from 200–900 kg/h of liquid chlorine

Figure 2.22 shows the transiting flows of chlorine, and the role of the superheater that maintains the chlorine in a gaseous state as it passes through *safety_valve_3* towards the production unit. The graph shows that the liquid and gaseous chlorine flows passing through the evaporator and entering the superheater are smooth; this is consistent with production requirements of 770 kg/h.

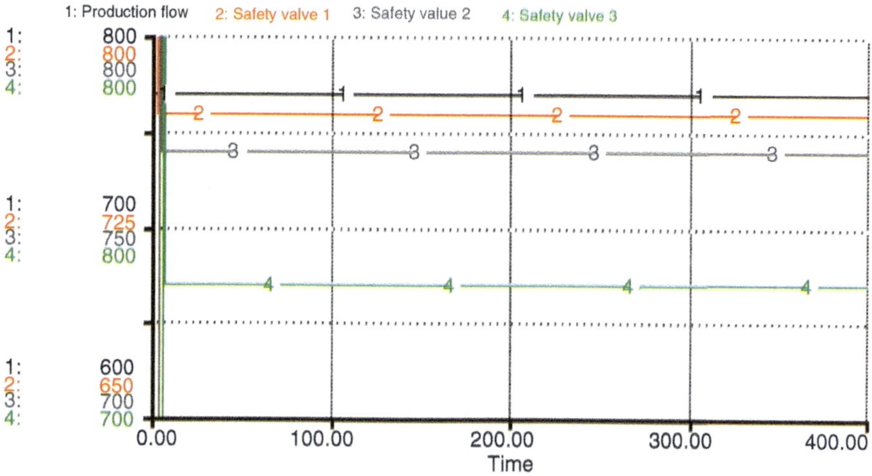

Fig. 2.21 Simulation of the flow of chlorine at the inlet and outlet of the evaporator and the outlet of the superheater. Fluctuations correspond to changes in demand over time (x-axis). 1 = production flow (ordinate = chlorine flow ranging from 600–800 kg/h). 2 = flow of liquid chlorine in safety valve 1 (ordinate = chlorine flow ranging from 650–800 kg/h). 3 = flow of chlorine gas into safety valve 2 (ordinate = chlorine flow ranging from 700–800 kg/h). 4 = flow of chlorine gas in safety valve 3 at the outlet of the superheater (ordinate = chlorine flow rate of 700–900 kg/h)

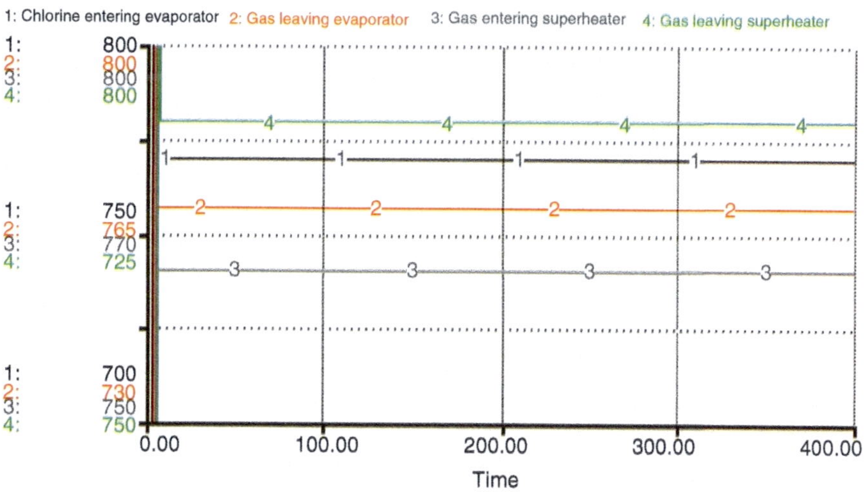

Fig. 2.22 Simulation of the flow of liquid chlorine entering the evaporator and chlorine gas that leaves the evaporator and superheater. 1 = liquid chlorine entering the evaporator (ordinate = chlorine flow rate 700–800 kg/h). 2 = chlorine gas at the outlet of the evaporator (ordinate = chlorine flow 730–800 kg/h). 3 = chlorine gas entering the superheater (ordinate = chlorine flow 750–800 kg/h). 4 = chlorine gas at the outlet of the superheater (ordinate = chlorine flow 650–800 kg/h)

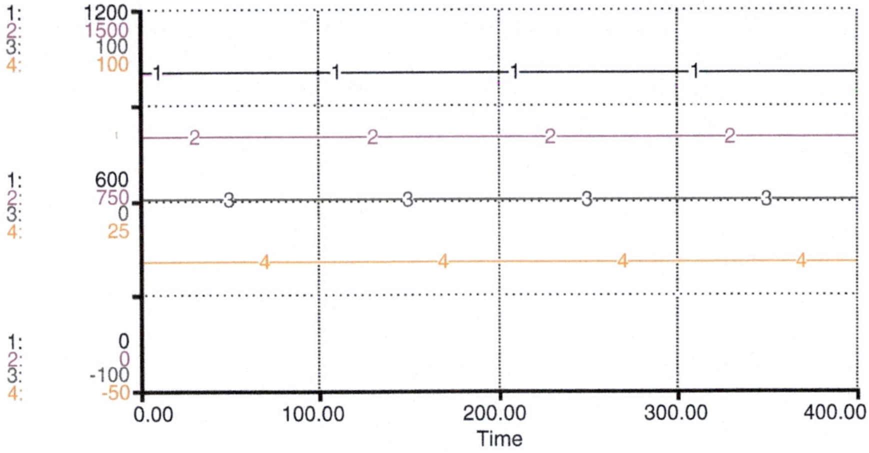

Fig. 2.23 State variables representing the hot water network that enables the chlorine state change. The system's status is assumed to be stable as the quantity of chlorine that is delivered is equal to that requested. 1 = hot water tank 1 (ordinate is the quantity of hot water 0–1200 kg). 2 = hot water tank 2 (ordinate the quantity of hot water 0–1500 kg). 3 = outflow of hot water from tank 1. 4 = outflow of hot water from tank 2

Figure 2.23 shows the state variables that regulate the supply of hot water to the evaporator and the superheater for the chlorine state change. It can be seen that the system is stable and that the heat supplied to the system for the chlorine state change is sufficient to ensure a steady flow of chlorine gas to the production unit.

This model of the discharge and transport of chlorine is designed to illustrate the normal mode of operation (Fig. 2.24). It therefore provides a baseline for analysing failures and prevention and protection measures, which are discussed in the following sections.

2.3 Application to the Dynamic Analysis of Failure and Systemic Safety Measures

The model of the discharge and chlorine transfer system described above, only considered normal operating conditions. This model provides can be used as a baseline for studying potential system failures. This section outlines a method for analysing and modelling system failures using the examples of a small, and a large chlorine leakage in different parts of the system. The results are presented and discussed.

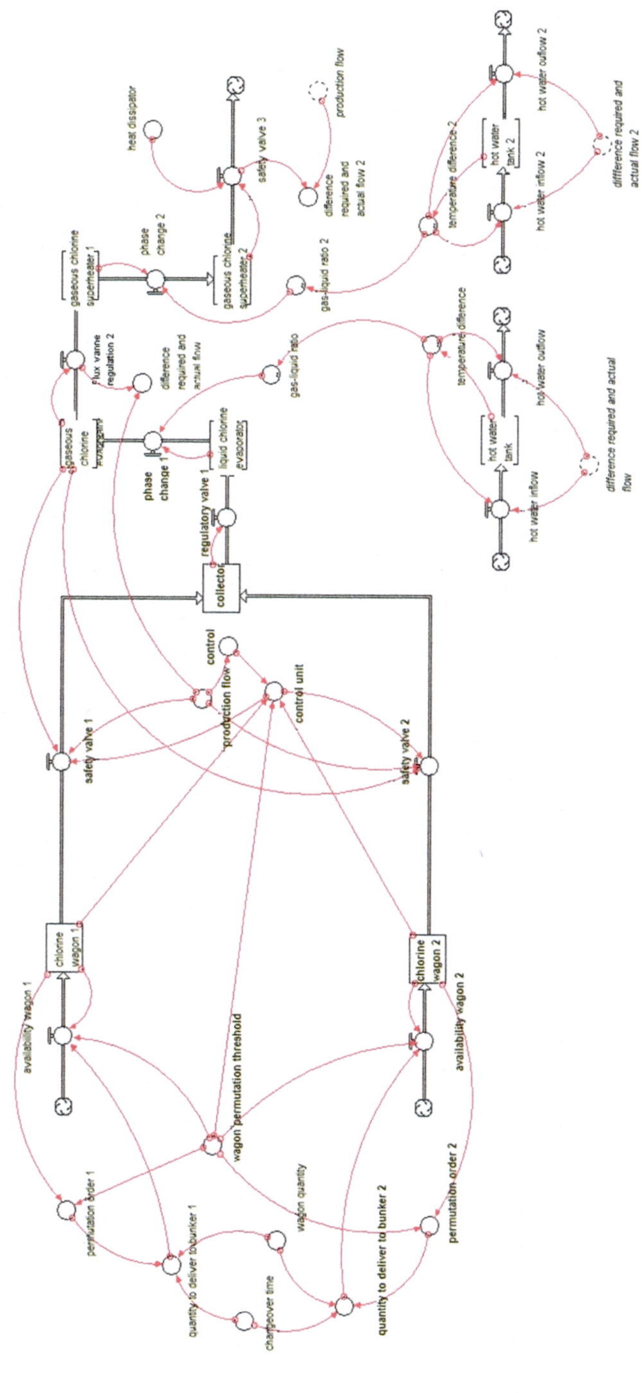

Fig. 2.24 General model of the discharge and transport of chlorine to the production unit

Table 2.6 Extract from French regulations governing the 'Use or storage of chlorine' (MEDAD 2007)

Use or storage of chlorine	
1. The total amount that may be present in the installation is greater than or equal to 25 t	AS
2. The total amount that may be present in the installation is greater than or equal to 1 t, but less than 25 t	A
3. In containers of a capacity greater than or equal to 60 kg, the total amount that may be present in the installation is greater than or equal to 60 kg but less than 1 t	A
4. In containers of a capacity of less than 60 kg, the total amount that may be present in the installation is:	
(a) Greater than or equal to 500 kg, but less than 1 t	A
(b) Greater than or equal to 100 kg but less than 500 kg	DC

The site that is the focus of the case study stores ≤25 tons of liquid chlorine
AS Authorisation with easement, *A* Authorisation, *DC* Declaration with periodic control

2.3.1 *Analysis of System Failures*

As noted earlier, like any industry that uses large quantities of chlorine, this facility has to comply with existing regulations. Under French law, any company that uses or stores 25 tonnes (or more) of chlorine is subject to specific *Use or storage of chlorine* regulations: this site typically stores around 60 tonnes of chlorine (Table 2.6).

Under the regulations, the site operator must implement a risk analysis, and evaluate accident scenarios as part of a hazard study. Here, system failures are analysed using the HAZard and OPerability Study (HAZOPS) method, which was developed in the United Kingdom in the early 1970s by Imperial Chemical Industries (ICI). The HAZOPS method can be applied in the design, construction and operational phases of industrial systems. It focuses on potential failures related to the physico-chemical properties of materials, and their consequences (Andrews and Moss 2002). The failures are often found in thermo-hydraulic circuits and measured in terms of quantities such as pressure, temperature and flow (Fig. 2.25).

The HAZOPS method, as applied in the case study, comprised the following steps:

1. Dividing the process or system into nodes that correspond to a particular step in the process. Three nodes were identified: the discharge bay (1), the chlorine line (2), and heat production (3);
2. Identifying the particular components involved (e.g., chlorine wagons, safety valves);
3. Identifying parameters related to the components (temperature, pressure, flow, etc.) and identifying potential failures using appropriate keywords (too high or low flow rate, high or low pressure, late or early triggering, etc.);
4. Examining possible causes of failures (pipe rupture, power failure, etc.);
5. Defining possible barriers to prevent or reduce the probability that a particular failure occurs (changes to hardware or processes, preventive maintenance, etc.);

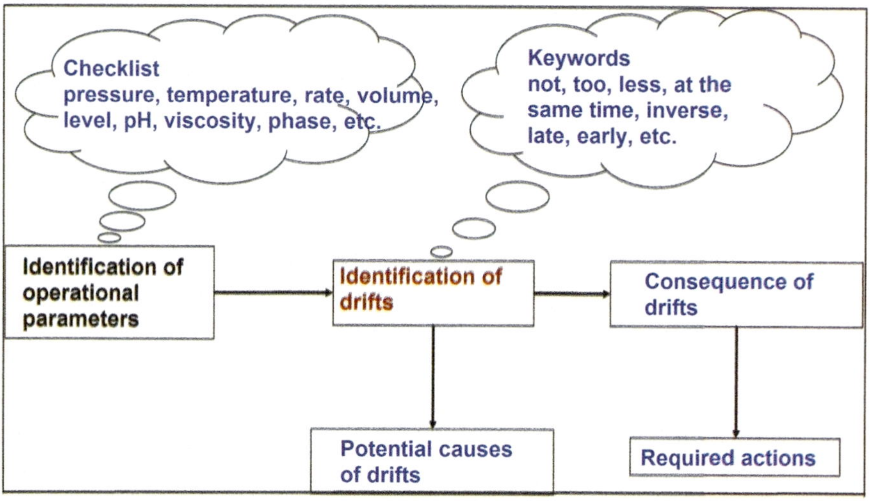

Fig. 2.25 Implementation of the HAZOPS method

Table 2.7 Steps in the HAZOPS method

Node	Component	Parameter and deviation	Cause of the deviation	Prevention barrier	Consequences	Protection barrier

6. Examining the effects, and quantifying the severity of the event in terms of their impact on health, environment and operations;
7. Identifying protective barriers that prevent or limit the effects on exposed areas.

These steps are formulated in a way that facilitates discussion and analysis by experts (see Table 2.7).

This method can also be adapted to the needs of the experts who use it, e.g., by adding an extra column to Table 2.7 to allow for further comments. It could also be supplemented by an evaluation of the probability of the occurrence of causes if the experts have access to the appropriate data about the reliability of the components involved. The joint use of probabilities and severity makes it possible to construct a two-way table (probability of occurrence of the event × severity of consequences) for use by the site operator, while the legislature provides a framework for the acceptability of the activity (in this case study, the use of chlorine for the synthesis of plastics).

From a practical point of view, the HAZOPS risk analysis involves three people:

- An independent expert who can guarantee that the analysis is carried out correctly;
- A representative of the company who is very familiar with the industrial process;

- A representative of the company who is responsible for operations.

If circumstances require, it is also possible to draw upon:

- One or more maintenance engineers or experts (electricity, corrosion, instrumentation etc.);
- A safety manager.

On average, it takes approximately one day to carry out a HAZOPS analysis for a particular system (e.g., the chlorine line, discharge unit, and hear production). This varies, however, depending on the complexity of the process and the components involved.

2.3.2 *Modelling and Dynamic Analysis of System Failures Using Stella®*

The normal operating conditions model of the discharge and chlorine transfer system was modified to incorporate unwanted events as a basis for analysing system failures. The following sections outline the steps and results of modelling these unwanted events that are seen as causes of deviation from normal operations. In each case, the consequences are evaluated without initially considering control mechanisms, alerts and the return to normal operating conditions.

Prevention and protection barriers are proposed and modelled in STELLA® to simulate their effect on system operation and determine their effectiveness. No attempt is made to present the full range of available prevention and protection measures, as this would substantially increase the length of this work. The primary purpose here is to present a formal framework for modelling risk management measures as way of validating their relevance in the design phase through simulating the behaviour of an industrial system.

2.3.2.1 Modelling a Leak of Liquid Chlorine During Discharge

Chlorine discharge incurs certain risks, notably for staff who carry out operations such as installing the chlorine wagon in the bunker, connecting the wagon to the pipeline for the transfer of liquid chlorine, and performing checks to ensure that the discharge is running smoothly. As the discharge is performed in the enclosed environment of the bunker, the risk that chlorine enters the environment and affects the surrounding population is negligible. The proposed scenario assumes that there is a chlorine leak in one bunker and some chlorine escapes into the surrounding atmosphere.

A leakage of liquid chlorine during discharge was modelled as follows in STELLA®:

Table 2.8 Variables used to model the chlorine leak at the collector

Modelling a leak of liquid chlorine during discharge		
Variable name	Definition	Function
liquid_ chlorine_leak	*collector* × 0.01	Determines the percentage of chlorine that escapes from the pipeline before it enters the evaporator.
total_ chlorine_loss	Stock	Total chlorine loss.
flow_valve_1	*collector* – *liquid_ chlorine_leak*	Chlorine flow rate arriving at the evaporator.

- A flow variable *liquid_chlorine_leak* that determines the quantity of chlorine escaping from the pipeline running to the evaporator. In the following example, the leak represents 1% of the normal flow, i.e., 7.7 kg of chlorine per hour (a slow leak).

Definition of the variable *liquid_chlorine_leak*
collector x 0.01

- A stock variable *total_chlorine_loss* that measures the total quantity of lost chlorine.

The introduction of this leak has an impact on the variable *regulation_valve_ flow_1*, which represents the flow of chlorine reaching the evaporator. Table 2.8 shows the variables that are used to model the chlorine leak.

Definition of the variable *flow_valve_1*
collector – *liquid_chlorine_leak*

Figure 2.26 shows the model that incorporates the chlorine leak in the STELLA® user interface.

The behaviour of the system when there is a leak of liquid chlorine in the two bunkers over a period of 400 h (the duration of the simulation) is shown in Fig. 2.27. This accident scenario assumes that there is no technical or human intervention to stop the discharge and transfer of chlorine: this would not be the case in practice. The bunkers are fitted with automatic safety devices, which activate the closure of safety valves, when the atmospheric chlorine sensor is triggered. Operators are only expected to intervene when the automatic systems fail, and this provides an additional organizational barrier. Figure 2.27 shows the following variables:

- The liquid chlorine leak that simulates the flow of chlorine escaping from the pipeline;

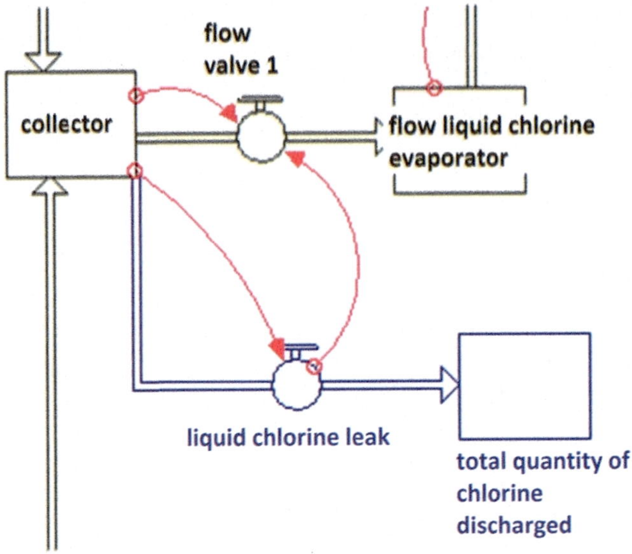

Fig. 2.26 Graphical representation of the model of the chlorine leak in the discharge unit in STELLA® software. The leak is modelled by two components: a flow variable that determines the percentage of liquid chlorine leaking from the pipeline, and a stock variable that indicates the total quantity of chlorine discharged into the bunker

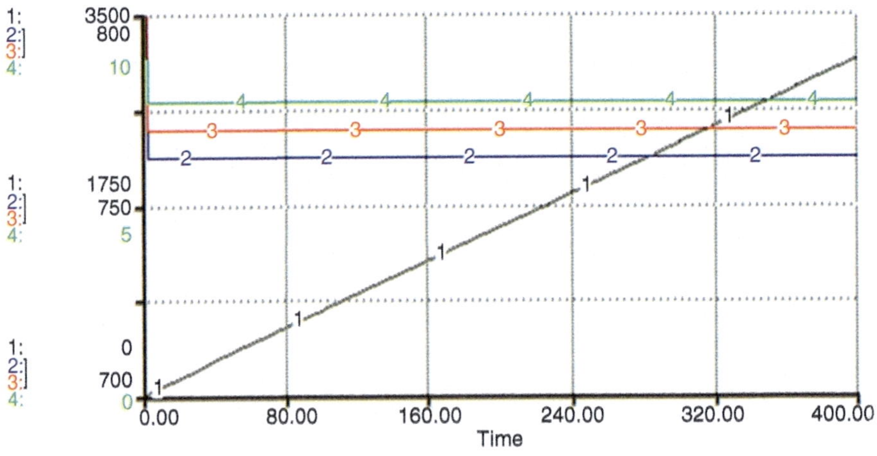

Fig. 2.27 Simulation of the effect of a small leak of liquid chlorine: 1 = quantity of chlorine released after 400 h (from 0–3500 kg); 2 = safety valve 1 (700–800 kg of chlorine per hour); 3 = the collector (700–800 kg of chlorine per hour); 4 = leakage of liquid chlorine, equal to 1% of the normal flow rate (0–10 kg of chlorine per hour)

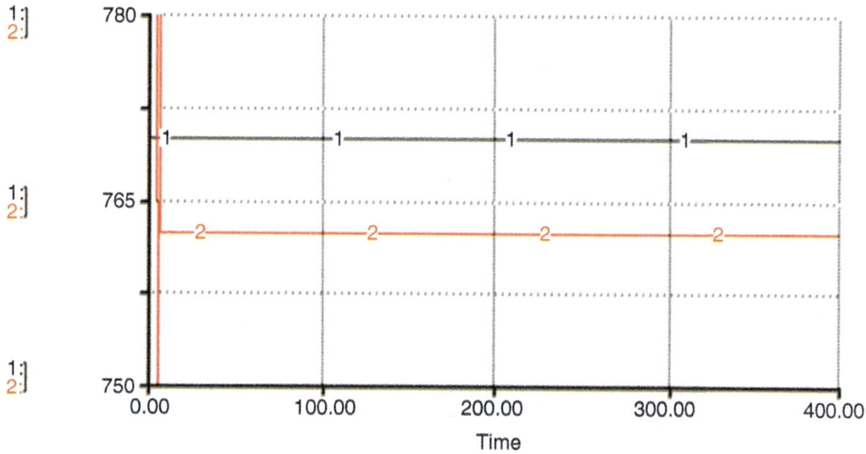

Fig. 2.28 Difference between the flow required by production and the quantity of chlorine gas passing through safety valve 3 (located just before the production unit). This figure shows that there is a small (but constant) difference between the quantity required and that received by the production unit (the y-axis represents the chlorine flow rate on a scale from 750 to 780 kg/h)

- The quantity of chlorine that flows through the collector and that remains stable with respect to demand;
- The quantity of liquid chlorine passing through safety valve 1 and which is routed to the evaporator. Logically this quantity is slightly less than that required by the production unit due to the upstream leakage that is introduced;
- The total quantity of chlorine that escapes from the pipeline and which remains at the bunker: about 3000 kg for 400 h of production.

Figure 2.28 shows the difference between the flow required by the production unit and the actual quantity of chlorine leaving the end of the line (at safety valve 3). The flow rate at the end of the line is 762.3 kg/h, a difference of 7.7 kg/h (i.e., 1%). This difference can be used later to implement an alert signal that warns the operators when there is a failure in the discharge unit.

The second scenario relates to a serious leakage of liquid chlorine during discharge, corresponding to 10% of the normal rate. This scenario does not take into account any intervention by operators. In this case, the model remains the same: only the value assigned to the leak rate represented by the variable *liquid_chlorine_leak* changes, as follows:

Definition of the variable *liquid_chlorine_leak*
collector x 0.1

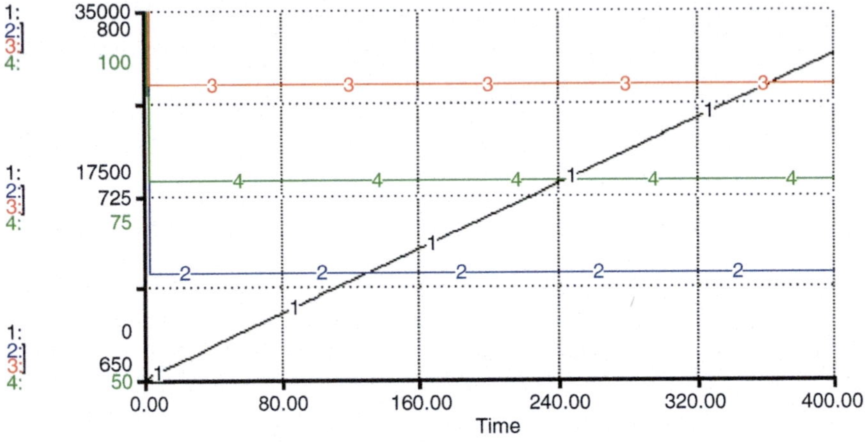

Fig. 2.29 Simulation of the effect of a serious leak of liquid chlorine: 1 = quantity of chlorine released after 400 h (0–35,000 kg of chlorine); 2 = safety valve 1 (650–800 kg of chlorine per hour); 3 = collector (650–800 kg of chlorine per hour); 4 = leakage of liquid chlorine equal to 10% of the normal flow rate (50–100 kg of chlorine per hour)

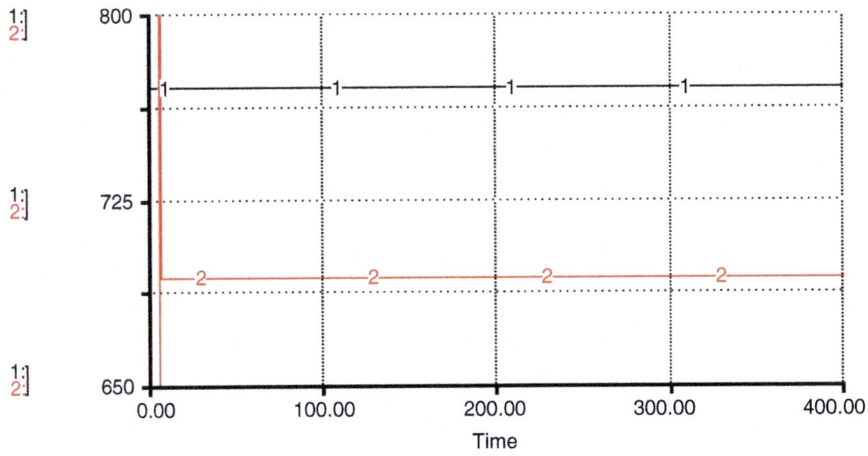

Fig. 2.30 Difference between the quantity requested by the production unit and the quantity of chlorine gas passing through safety valve 3 that is located just before the production unit (the y-axis represents the chlorine flow rate on a scale from 650,800 kg/h). 1 = production flow; 2 = flow through safety valve 3

The occurrence of serious chlorine leak of 77 kg/h (i.e., 10%) corresponds to a major release of chlorine into the bunker, exceeding 30,000 kg of chlorine after 400 h, as shown in Fig. 2.29.

Figure 2.30 shows the difference between the flow of chlorine gas at the input to the production unit and its demand: the difference is large enough to be recognised by the control room operators as an indication that there has been a leak. They can use this information to manually trigger an alert that leads to the closure of the chlorine transfer line.

The following section describes a scenario where the chlorine leak occurs at the end of the line, where chlorine gas reaches the production unit. This time, the leak is modelled as a total rupture of the pipeline.

2.3.2.2 Modelling a Leak of Chlorine Gas at the End of the Line

A leak of chlorine gas at the end of the line can be particularly dangerous because the pipe that carries the chlorine to the production unit is not protected by any special containment measures. Even a relatively small leak can cause the formation and diffusion of a toxic cloud of chlorine gas within the factory and over several metres beyond its perimeter.

In this case, there is a rupture of the pipeline at the outlet of safety valve 3, caused by a blow from a hoist., so all the chloring as escapes The simulation excludes the action of operators and protection systems. The rupture is modelled by the following variables:

- The flow variable *gaseous_chlorine_leak*, which simulates the rupture in the pipeline. This variable represents the total flow from the pipeline located after safety valve 3;

Definition of the variable *gaseous_chlorine_leak*
flow_safety_valve_ 3 x 1

- The stock variable *exterior_pipeline_chlorine*, which models the quantity of chlorine leaving safety valve 3;
- The flow variable *production_input*, which represents the section of the pipeline located after the rupture. This variable represents the flow of chlorine reaching the production unit taking into account the losses caused by the leak. In this scenario, the amount is zero;

Definition of the variable *production_input*
exterior_pipeline_chlorine – *gaseous_chlorine_leak*

- The stock variable *quantity_chlorine_lost*, which represents the total amount of chlorine released into the atmosphere during the simulation.

Table 2.9 lists the variables used and modified for the simulation of the rupture at the end of the pipeline.

Figure 2.31 shows the ruptured pipeline in the STELLA® interface.

Table 2.9 List of variables used to model a chlorine leak at the collector

Modelling a leak of chlorine gas at the end of the line		
Variable name	Definition	Function
gaseous_chlorine_leak	*flow_safety_valve_ 3* × 1	The pipeline rupture
exterior_pipeline_chlorine	Stock	The amount of chlorine that transits the exit of safety valve 3
production_input	*exterior_pipeline_chlorine* − *gaseous_chlorine_leak*	Part of the pipeline located after the break
quantity_chlorine_lost	Stock	Total amount of chlorine released into the atmosphere during the simulated period

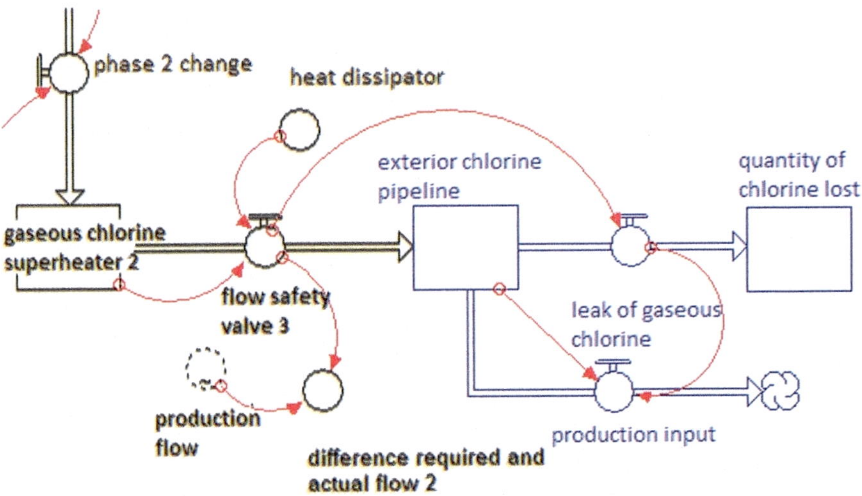

Fig. 2.31 Graphical representation of the model of the ruptured pipe at the outlet of safety valve 3 in STELLA®. The model of the rupture includes two flow variables that determine the flow of chlorine released into the atmosphere and the amount that reaches the production unit. In this example, the quantity reaching the production unit is zero

Figure 2.32 shows the behaviour of the system following the rupture of the pipe. The variables that are analysed relate to:

- The leak of chlorine gas and flows through safety valve 3; this is determined by production flows. The amount of chlorine released into the atmosphere is 770 kg/h (i.e., 100%).
- The flow of chlorine gas at the input to the production unit: the flow is 0 kg/h because the pipe located at the outlet of safety valve was completely severed. In other words, all the chlorine gas is being released into the atmosphere.

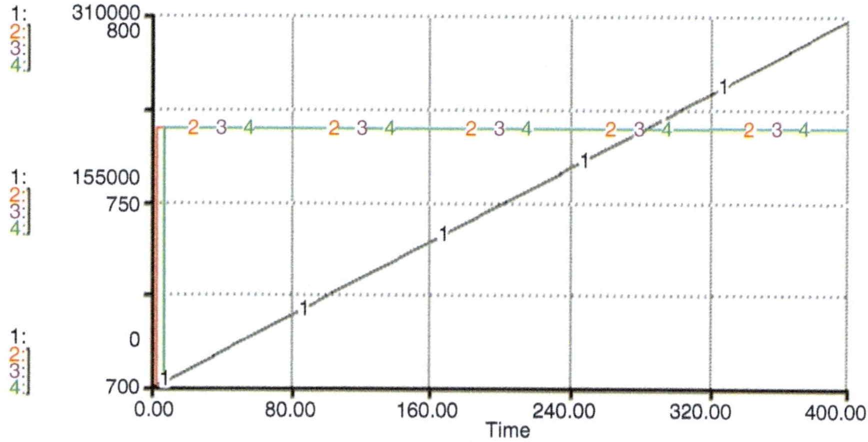

Fig. 2.32 Simulation of the ruptured pipe at the outlet of safety valve 3. Variable 1 (the quantity of chlorine gas this is released is between 0 and 340,000 kg, i.e., the quantity of chlorine released is the total quantity contained in the two wagons. The ordinates of the following variables correspond to the chlorine gas flow rate on a scale from 700 to 800 kg/h; 2 = chlorine gas flow passing through flow valve 1; 3 = chlorine gas flow passing through safety valve 3; 4 = leakage of chlorine gas

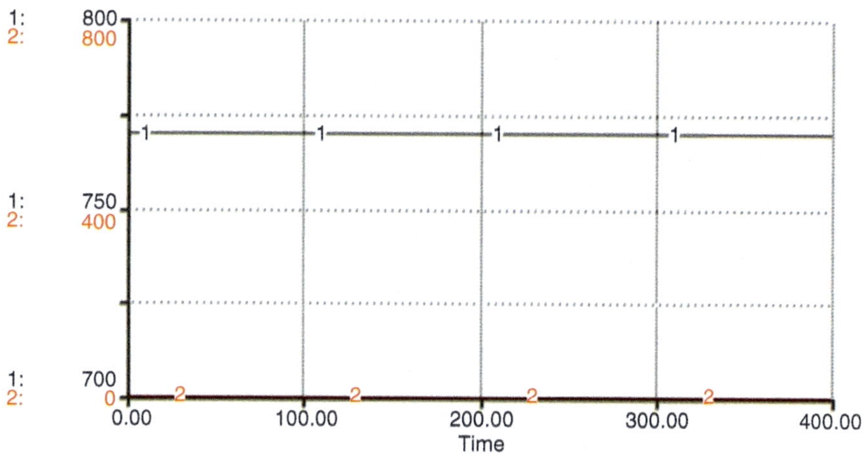

Fig. 2.33 Difference between the flow required by the production unit and the actual flow. 1 = demand for chlorine gas (700–800 kg/h); 2 = flow of chlorine gas arriving at the production unit (0–800 kg/h). The amount of chlorine reaching the production unit is 0

In this scenario, the amount of chlorine lost is equivalent to the quantity left in the two wagons, i.e., 300,000 kg. Figure 2.33 shows that no chlorine is reaching the production unit due to the rupture of the line.

2.3.2.3 Risk Analysis of the Discharge System and Chlorine Pipeline

This section presents the risk analysis for the different chlorine leaks modelled above. It takes account of the potential causes that can lead to unwanted events and the subsequent effects. On the basis of this analysis, prevention and protection measures can be proposed; these are modelled in detail later in the chapter.

The risk analysis is based on the formal identification of system failures using the HAZOPS method. The results are shown in Table 2.10, including the scenarios described above. The prevention and protection measures that are modelled later are shown in bold typeface. These technical and organizational measures are intended to change the structure of the monitoring system and to warn operators of a system malfunction. When there is a leak, these measures are activated, leading to operator intervention, and thereby triggering the appropriate contingency plans.

The outcomes of these scenarios are evaluated using the PHAST® software package, which can simulate the atmospheric dispersion of a chlorine release. PHAST® consists of several integrated digital models that take into account different types of release (release of toxic substances, the formation and evaporation of droplets and liquid spills) and dispersion (jet dispersion, integrated model of heavy gas dispersion and a Gaussian model of passive dispersal). PHAST® is widely used in the oil and process industries, and by state inspection services.

PHAST® was used to model how far the chlorine gas cloud would propagate in the atmosphere following a pipeline rupture, based on the time of the release, the flow rate, the air temperature, the wind speed, etc. These parameters are listed below, together with their assigned values in the scenarios:

- Product type: chlorine;
- Atmospheric parameters (the values are drawn from meteorological information services):
  - Wind speed: 3 m/s and 5 m/s and corresponding to the values used at the site;
  - Wind direction: the wind direction could not be predicted, so it was assumed that the release formed a circular cloud;
  - The height of the toxic cloud (1.5 m);
  - The physical relief of the outside environment. This was considered because of the height of the pipeline rupture (3 m) and the small number of tall buildings and trees near the chlorine line;
  - Cloud cover expressed as a percentage: the default setting of partially cloudy was chosen (50%);
  - The air temperature: 20 °C;
  - Atmospheric stability: six classes of stability are defined by meteorologists, D (neutral conditions) and F (moderately stable conditions) were selected as a function of wind speed and sunshine;
  - A temperature inversion with a defined height: this parameter was not included;
  - Humidity expressed as a percentage: 50% was selected.
- Release rate: a continuous release (i.e., 100%) was assumed at a rate of 770 kg/h.

Table 2.10 Example of the risk analysis of failures in the discharge system and the chlorine pipeline as a function of the chlorine leak scenarios that are discussed. Prevention and protection measures that are modelled (later in the chapter) are shown in bold, their influence on system behaviour was studied

Node	Component	Parameter and deviation	Cause of the deviation	Prevention barrier to implement	Potential consequences	Protection barrier to implement
Chlorine tank transfer unit	Outlet pipe from safety valve 1 or 2 (collector)	Reduced flow leaving the collector	Leak caused by badly joined pipes	Preventive maintenance to replace materials	Formation of a toxic cloud in one of the two bunkers	**Alarm**
						Closure of safety valves
				Inspection		
				Atmospheric chlorine sensors		**Automatic shutdown of the installation**
						Activation of internal accident procedures
			Leak caused by mishandling	Training of personnel	Formation of a toxic cloud in one of the two bunkers	**Alarm**
				Atmospheric chlorine sensors		**Closure of safety valves**
						Automatic shutdown of the installation
						Activation of internal accident procedures
			Malfunction of the heating system	Temperature control	Reduction or temporary halt in production	
				Regular monitoring and preventive maintenance		
			Malfunction of the regulatory valves	Checks of valve opening	Reduction or temporary halt in production	

For all of the atmospheric dispersion scenarios, the three chlorine concentrations that are considered are:

1. the Irreversible Effects Threshold (IET; 25 ppm);
2. the Initial Lethal Effects Threshold (ILET), which is fatal for 1% of the exposed population (160 ppm);
3. and the Significant Lethal Effects Threshold (SLET), which is fatal for 5% of exposed persons (183 ppm).

These thresholds correspond to 30 min of exposure. The calculation of the area affected by the release was based on selected data from a report by the Ministry of Ecology and Sustainable Development on irreversible lethal effects affecting 1% of the exposed population, and an April 2005 report on significant lethal effects affecting 5% of the exposed population:

- For the IET: C2.3 * t = a constant passing through the point 25 ppm for an exposure time of 30 min.
- For the ILET: C1.9 * t = a constant passing through the point 160 ppm for an exposure time of 30 min;
- For the SLET: C1.9 * t = a constant passing through the point 183 ppm for an exposure time of 30 min;

The consequences are modelled as shown in Table 2.11, taking into account the thresholds for lethal and irreversible effects on workers and the general population beyond the site boundaries.

The results of the simulation of the atmospheric dispersion of chlorine are shown in Table 2.11, using the values defined above. The hazard distances calculated using PHAST® do not exceed the physical boundaries of the industrial site for the two lethal effects thresholds.

Table 2.11 shows that only scenarios corresponding to the IET will result in a somewhat toxic cloud of chlorine extending beyond the site's perimeter. For the IET scenario, the dispersion of this toxic cloud, given a pipeline rupture at a flow rate of 770 kg/h is shown in Fig. 2.34. The mapping was performed using PHAST®.

The results of these simulations of atmospheric chlorine dispersion emphasize that the exposure of the local population in these scenarios is low. In order to protect the health of the workforce, the local population, the environment and production facilities, however, it is essential to define prevention and protection measures.

Table 2.11 Chlorine cloud propagation distances calculated at wind speeds of 3 and 5 m/s for IET, IELT and SLET scenarios. ND = Not Determined for the wind speed used. These results show that only scenarios related to the Irreversible Effects threshold will have an impact beyond the physical limits of the facility

Chlorine release at 770 kg/h		
	Wind speed 3 m/s	Wind speed 5 m/s
IET	380 m	114 m
ILET	75 m	ND
SLET	64 m	ND

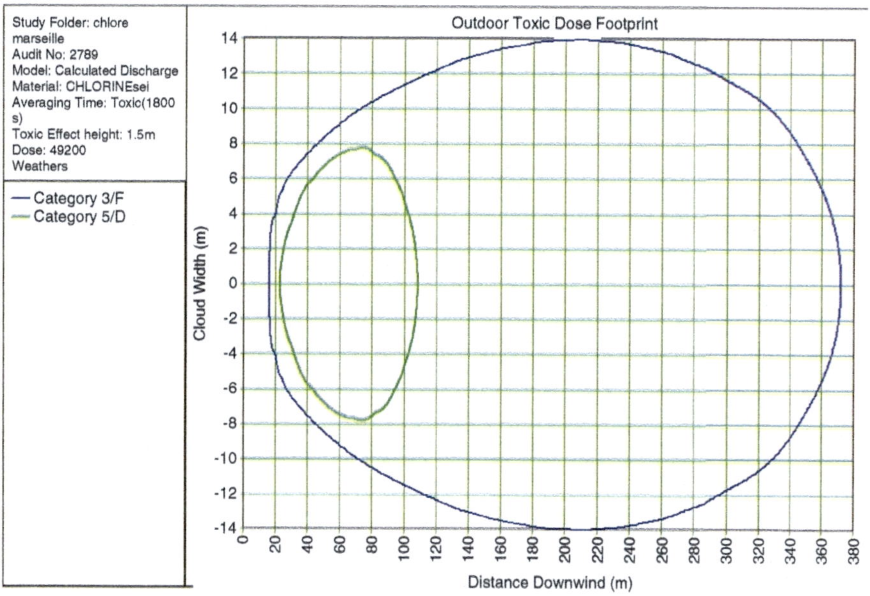

Fig. 2.34 Mapping of the dispersion of a cloud of chlorine at a concentration of 25 ppm (IET) for wind speeds of 3 and 5 m/s

These measures, which were identified in the risk analysis (Table 2.10), were incorporated into the model. The following sections present the results of the modelling and simulation of these measures.

2.3.2.4 Modelling of Preventive and Protection Measures

The first protection measure that will be considered is an alarm system. Alarms can be activated manually, when the operators detect a leak, or automatically when the difference between the demand for chlorine and the actual flow rate exceeds some limit (e.g., 1%). This type of low-pressure safety device is widely used in process engineering. Figure 2.39 shows the STELLA® software model of a chlorine leakage at the end of the line, following the rupture of the pipeline. To make the event more realistic, a few changes were made to take into account the time of the leak, the alert system and how long it would take to repair the pipeline. The revised model includes the following variables:

- The auxiliary variable *leak_part*, which is the percentage of chlorine that escapes into the atmosphere. This amount can be modelled as a random value but, in this case, the line is completely severed, so the variable is set to 1, i.e., 100% of the chlorine flow.

- The auxiliary variable *leak_time*, which is the time when the leak occurs. This can be modelled by a random function, to simulate the unpredictability of the event. In this case *leak_time* is set to 200 h, and the random function is set to zero.

Definition of the variable *leak_time*
(RANDOM (1, 400)) x 0 + 200

- The auxiliary variable *leak_interval*, which is the period between two leaks on the same component. If a leak appears, temporary repairs may be needed to bring the pipeline back into service, pending replacement of the failed part. Another event (e.g., a leak caused by poor welding or joints) may occur before the part is replaced. This time was arbitrarily set at 72 h after the system shuts down.
- The auxiliary variable *leak_rate* represents the amount of chlorine that is released in a given moment. The value will be zero when the system is operating normally. It will become non-zero after a repair has been carried out if another leak occurs. The PULSE function takes into account the preceding variables in order to model the leak.

Definition of the variable *leak_rate*
PULSE (*leak_part*, *leak_time*, *leak_interval*)

- The flow variable *gaseous_chlorine_leak* was amended as follows to reflect the parameters of the leak rate:

Definition of the variable *gaseous_chlorine_leak*
safety_valve_flow_3 x *leak_rate*

- The auxiliary variable *alert* represents a chlorine leak protection measure. This variable is activated when the difference between the flow of chlorine reaching the production unit and the expected flow exceeds ±10%.

Definition of the variable *alert*
IF (*production_input* < (*production_flow* × 0.99) OR *production_input* > (*production_flow* × 1.10)) THEN 1 ELSE 0

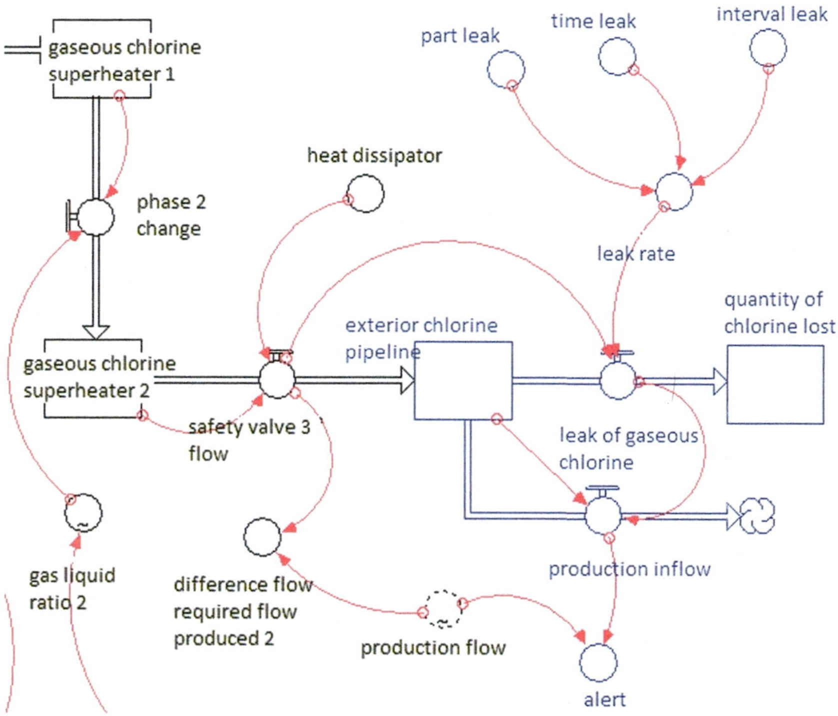

Fig. 2.35 Modelling a chlorine leak at the end of the line and the warning system

Figure 2.35 shows the variables that were used to model the leak at the end of pipeline, including the leak alert system. Stock variables and the flow variable *production_input* remain the same.

The warning system is an initial safety measure that prompts the operators to intervene. In the case of chlorine leak, the system must be shut down safely. The shutdown and the repairs that are carried out are modelled by the following variables:

- The flow variable *alarm_trigger* is activated as soon as the alert is given: it is equal to the variable *Alert* and therefore takes the value 1.
- The stock variable *alarm* is activated when the alarm is triggered: its initial value is 1 and this changes to 0 when the alarm is disabled.
- The auxiliary variable *alert_state* is activated when the alarm sounds. This variable interacts with the *control* variable triggering the closure of safety valves and system shutdown.

Definition of the variable *alert_state*
IF *alarm* THEN 1 ELSE 0

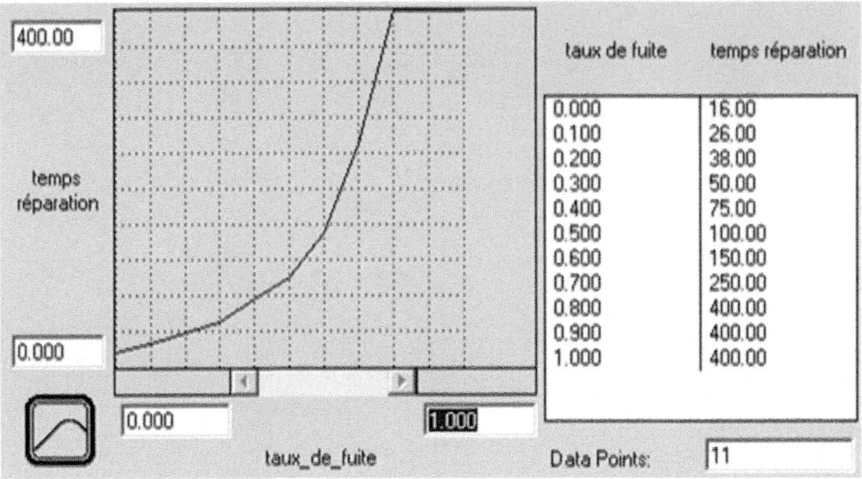

Fig. 2.36 Relation between the leak rate, which here corresponds to the severity of the failure and the time to repair the system. In the case of a total pipeline rupture, the time needed to repair of the system is 400 h

- The auxiliary variable *control* is modified by the introduction of the alert state. Its initial values relate to the regulation of the heating system and chlorine demand, but it also includes the alert conditions that lead to the activation of the control station in order to close safety valves when a leak occurs.

Definition of the variable *control*
IF (*production_flow* > 770 OR *production_flow* < 770) THEN 1 ELSE (IF *alert_state* = 1 THEN 1 ELSE 0)

When the alert is triggered, the system must be shut down, and then repaired to restore it to its normal operating state. Once these repairs have been made, the alert is suspended. Modelling this sequence of actions requires the following variables:

- The auxiliary variable *repair_time* for the leak is non-linearly proportional to the percentage of gas emitted into the atmosphere (the leak rate). The higher the leak rate, the longer the repair will take (measured in hours) (Fig. 2.36). Once this rate exceeds a certain level (e.g., when the pipeline ruptures) the repair time exceeds the duration of the simulation (400 h).
- The time for the repair and the alert interact with the auxiliary variable *alarm_all_clear*, which represents the time needed to end the alarm.

Definition of the variable *alarm_all_clear*
DELAY (*Alert*, *repair_time*)

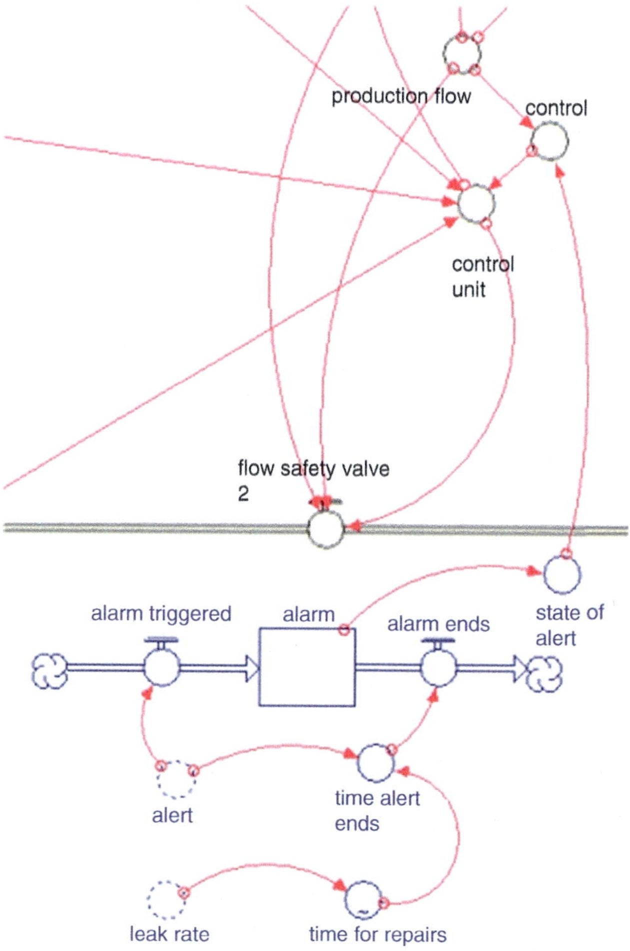

Fig. 2.37 Model of the alarm trigger and system shutdown

- The flow variable *alarm_stop* is activated by the *alarm_all_clear* variable: it represents the time when the system can resume normal operations following the completion of repairs.

Figure 2.37 shows the model of the alert and system shutdown following a rupture at the end of the chlorine pipeline. This same model is used for a (small) leak at the collector located at the output of the chlorine wagon discharge bunkers.

The analysis of the system in abnormal conditions (i.e., after the failure) highlighted the need to install sensors to monitor atmospheric chlorine levels. These sensors can detect even small chlorine leaks in open and confined environments.

The analysis showed that a chlorine sensor could be installed at the collector (in the confined environment). This sensor was modelled as follows:

- The auxiliary variable *sensor*, which measures the quantities of chlorine as the equivalent of 1 l in the area concerned. At a leak rate of 0.5% per hour in the collector, a very small leak, it takes about 1 h for the sensor to detect the leak and trigger the alarm.

Definition of the variable *sensor*
IF *quantity_chlorine_not_in_collector* > 1 THEN 1 ELSE 0

- It is also necessary to define a flow variable that corresponds to the chlorine evacuation variable (e.g., via aspiration through a deposition column) while the collector is being repaired. The flow variable *chlorine_evacuation* is activated when the alarm state is activated, i.e., when the alarm is triggered. The system is then shut down safely and repairs can begin.

Definition of the variable *chlorine_evacuation*
IF *alert_state* = 0 THEN 0 ELSE *quantity_chlorine_not_in_collector*

- The auxiliary variable *leak_duration* represents the time the leak occurred; this was arbitrarily set at 106 h. The leak is detected when the quantity of chlorine in the atmosphere exceeds 1 l, which activates the sensor that triggers the alert.
- The auxiliary variable *collector_leak* is defined as follows to take into account the duration of the leak:

Definition of the variable *collector_leak*
PULSE (*leak_duration*, 100)

Figure 2.38 shows the relationships between the variables used to simulate a small leak in the collector and the installation of an atmospheric chlorine sensor.

The model was revised as follows to include the characteristics of the leak at the collector:

- The alert is defined as follows, to take account of information from the sensor, and variations in the chlorine flow rate:

Definition of the variable *alert*
IF (*production_input* < (*production_flow* × 0.99) OR *production_input* > (*production_flow* × 1.10)) THEN 1 ELSE (IF *sensor* = 1 THEN 1 ELSE 0)

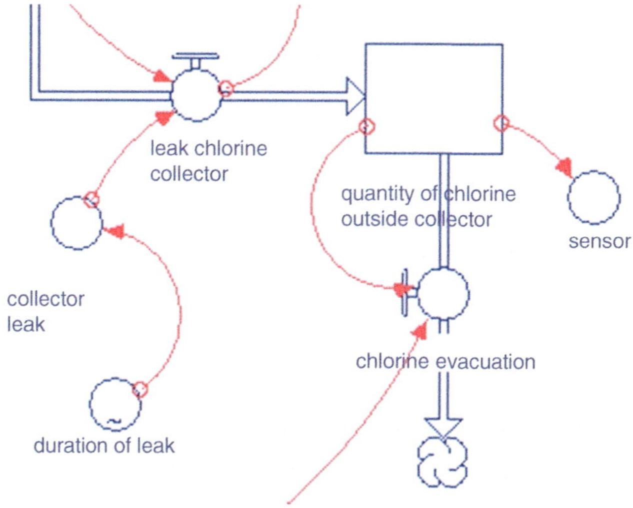

Fig. 2.38 Model of a small chlorine leak at the collector and the installation of a chlorine sensor

- The auxiliary variable *restart* was introduced to simulate the return to normal and the resumption of discharge operations and the transport of chlorine. The various units resume their operations when the alarm is stopped.

Definition of the variable *restart*
IF *alarm_stop* = 1 THEN 0 ELSE 1

- The auxiliary variable *alert_state* was modified as follows to take into account the information from the alarm and the restart. This variable interacts with both the auxiliary variable *control* and the flow variable *chlorine_evacuation*.

Definition of the variable *alert_state*
IF *alarm* x *restart* > 0 THEN 1 ELSE 0

Figure 2.39 shows the modifications that were made to the alert and its effects on system operation.

Table 2.12 shows all the variables that were used to simulate the behaviour of the industrial system including the additional prevention and protection measures.

The next section examines the impact of the protection and prevention barriers, introduced above, on the behaviour of the system in order to assess the effectiveness of the Important Safety Elements (ISE).

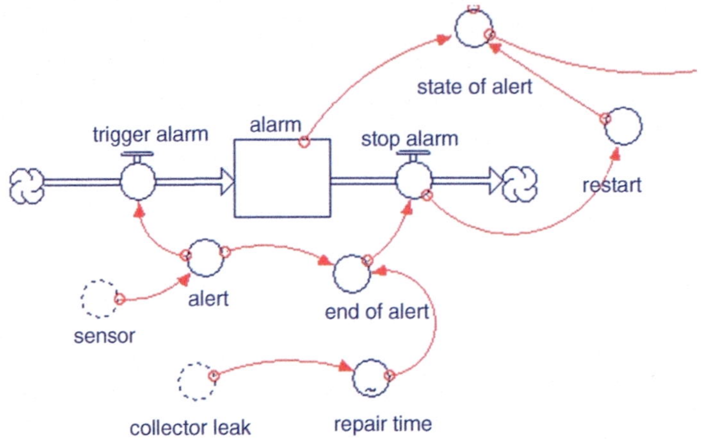

Fig. 2.39 Model of the alarm, the alert and the restart of the discharge system and chlorine transport

2.3.2.5 Simulation of the System with Prevention and Protection Measures

Following the introduction of the ISE the system's behaviour is analysed from two perspectives: the severing of the end pipeline causing a long-term system shutdown; and a small leak in the collector that leads to a temporary system shutdown while repairs are carried out.

Severing of the End of Pipeline

When the pipeline is severed at time $t = 200$ h, it has two effects:

1. The flow rate of chlorine entering the production unit falls to 0, which is reflected in a drop in the pressure in the pipeline;
2. The alarm is activated due to the change in pressure.

Figure 2.40 shows the difference between the required and actual flows at the evaporator outlet (1), and the difference between these flows at the superheater outlet (2). When the system is operating normally, the required and actual rates should be the same at both points. When the pipe is severed (t = 200 h) the difference between the required and actual flows of chlorine instantly rises to 770 kg/h. This activates the alert that triggers the shutdown of the facility. This safety measure is a negative feedback loop as it limits the activity of the system.

The activation of the alert triggers the alarm that, in turn, activates an alert message corresponding to the *Control* variable. This message is sent to the operator control station that proceeds to close safety valves 1 and 2 in the discharge bunkers.

Table 2.12 Variables representing prevention and protection measures

Modelling of prevention and protection barriers		
Variable name	Definition	Function
leak_part	Stock	Percentage of chlorine that escapes into the atmosphere
leak_time	(RANDOM (1400)) × 200 + 0	Time that the leak occurs
leak_interval	Set to 72 h	Time that elapses between two leaks on the same element
leak_rate	PULSE (*leak_part*, *leak_time*, *leak_interval*	Quantity of chlorine this is released at one time and may recur some time after the initial leak
gaseous_chlorine_leak	*flow_safety_valve_3* × *leak_rate*	Rate of chlorine release
alert	IF (*production_input* < (*production_flow* × 0.99) OR *production_input* > (*production_flow* × 1.10)) THEN 1 ELSE 0	Activated when the flow of chlorine to the production unit is either <99% or > 1% of expected flows
alert_state	IF *alarm* > 0 THEN 1 ELSE 0	Activated when the alarm sounds
control	IF (*production_flow* > 770 OR *production_flow* < 770) THEN 1 ELSE (IF *alert_state* = 1 THEN 1 ELSE 0)	Modified by the introduction of the alert state
repair_time	Graph	Non-linearly proportional to the percentage of gas leaking.
alarm_all_clear	DELAY (*alert*, *repair_time*)	Time required to raise the alarm
alarm_stop	Binary: 0 or 1	Activated by the variable alarm_all_clear
sensor	IF *quantity_chlorine_not_in_collector* > 1 THEN 1 ELSE 0	The amount of chlorine measured as the equivalent of 1 litre in the space in question/=
chlorine_evacuation	IF *alert_state* = 0 THEN 0 ELSE *quantity_chlorine_not_in_collector*	Chlorine evacuation (by aspiration to a ventilation chimney, for example) while the collector is undergoing repairs
leak_duration	Set at 106 h	The time the leak occurs.
collector_leak	PULSE (*leak_duration*, 100)	Appearance of the chlorine leak
restart	IF *alarm_stop* = 1 THEN 0 ELSE 1	The return to normal operations
alert_state (modified)	IF *alarm* x *restart* > 0 THEN 1 ELSE 0	System restart

Figure 2.41 shows that the discharge of the chlorine wagons being halted at $t = 200$ h. The shutdown lasts for the remainder of the simulation because major repairs are needed to the pipeline: the variable *repair_time* (defined above) assumes that the system will be shut down for 400 h when the pipeline is severed (the flow drops to 0).

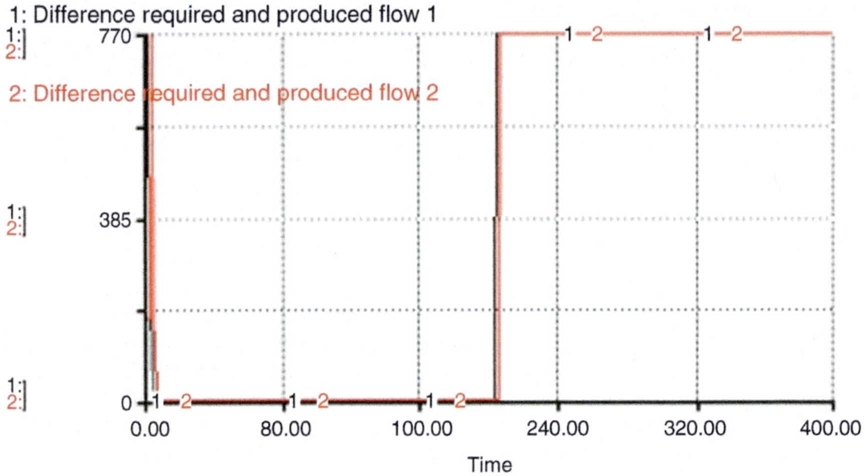

Fig. 2.40 Difference between required and actual flow at the outlet of the evaporator and the superheater. If this difference exceeds 1% (as is the case here) the alert is triggered in the event of a leak in the chlorine line

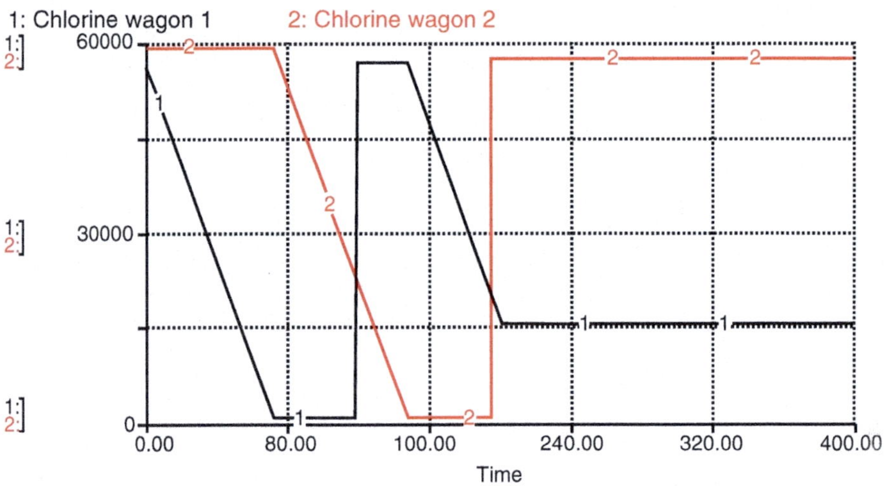

Fig. 2.41 Shutdown of chlorine wagon discharge following the trigger of an alert due to the severing of the end of the chlorine pipeline at time t = 200 h. The system remains shut down for the duration of the simulation

The shutdown of the discharge system cuts off the chlorine supply to the entire facility, first to the collector and control valve 1 (Fig. 2.42), and then to the evaporator and superheater (Fig. 2.43).

The system's behaviour following the severing of the pipeline demonstrates the critical role of the prevention and protection measures. In practice, the activation of

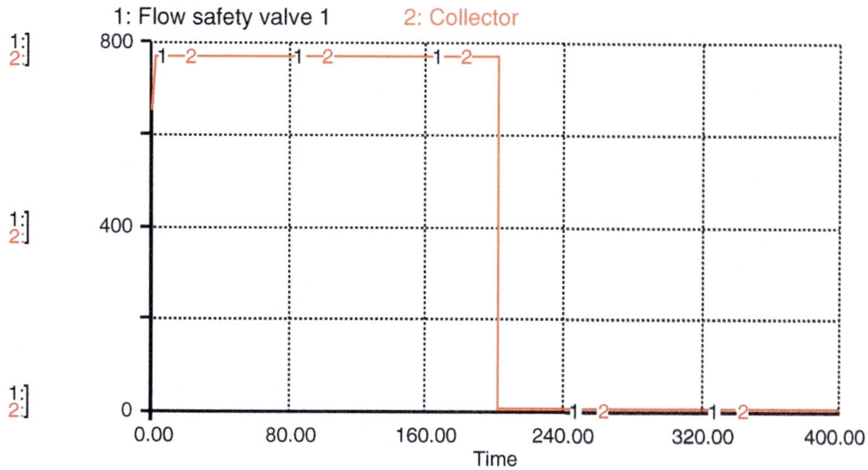

Fig. 2.42 Chlorine flow to the collector and control valve 1 in normal mode and after the shutdown of the system for safety reasons

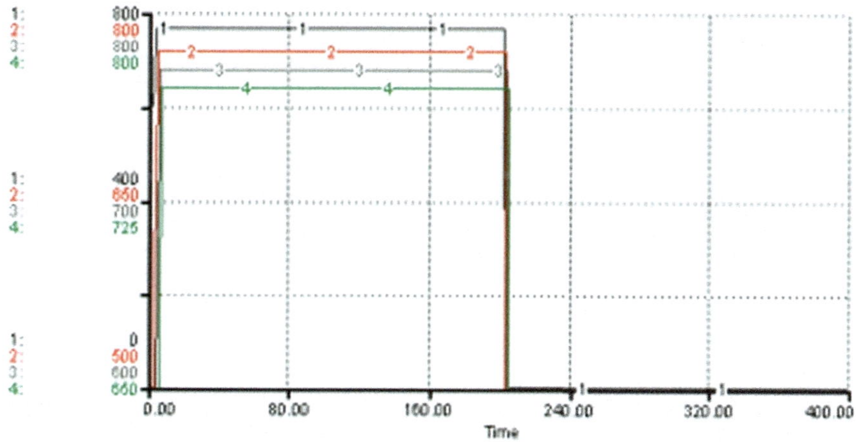

Fig. 2.43 Chlorine flow to the evaporator and superheater in normal mode and after the shutdown of the system for safety reasons.1 = chlorine flow to evaporator 1; 2 = chlorine flow to evaporator 2; 3 = chlorine flow to superheater 1; 4 = chlorine flow to superheater 2

safety valves rapidly brings the system to a safe state. This prevents the emission of pressurized chlorine gas into the atmosphere. The model provides an initial validation of the safety measures in the chlorine discharge unit. The following section highlights the role of atmospheric chlorine sensors at the collector, in the enclosed environment of the discharge bay.

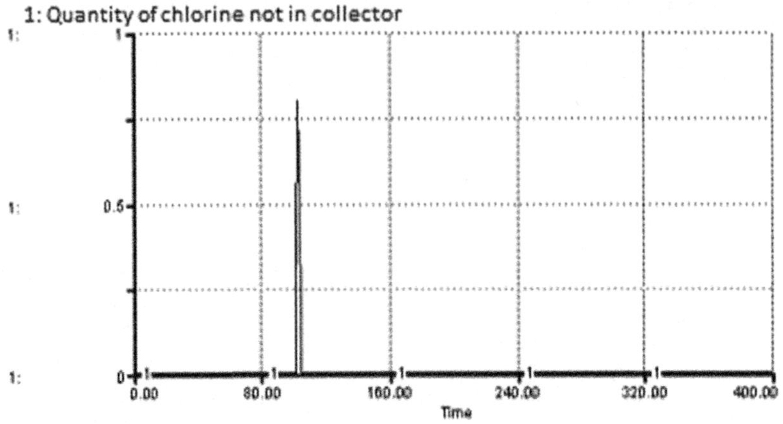

Fig. 2.44 Simulation of a small chlorine leak at the outlet of the collector and the activation of the sensor once 1 l of chlorine has been released

A Small Leak at the Collector

The occurrence of a very small leak at the collector (of the order of 0.5% of flow) does not trigger an alert, as the detection system's threshold is 1% of flow. The chlorine sensor, however, can be activated when the quantity of chlorine released into one of the two bunkers exceeds the threshold of 1 l.

Figure 2.44 shows the simulation of a 0.5% chlorine leak at the outlet of the collector, starting at $t = 100$ h. This figure shows that it takes about 1 h for the amount of chlorine that is released to reach the threshold of 1 l, which activates the atmospheric chlorine sensor.

As with the simulated chlorine leak at the end of the line, as soon as the sensor is activated it triggers an alert that in turn triggers the chlorine leak alarm. The latter stops the discharge of chlorine wagon 2 (Fig. 2.45).

The system will be shut down for the time it takes to carry out repairs (about 15 h). Once the system is considered safe to begin operating again, normal operations can be resumed.

Figure 2.46 shows the flow of chlorine passing through safety valves 1 and 2: it highlights the effectiveness of the chlorine sensor and the sequence of responses to its activation that lead to the temporary shutdown of discharge activities for about 15 h.

Another consequence of this shutdown is the cessation of the chlorine flow to the collector and to flow safety valve 1 (Fig. 2.47).

The shutdown of the discharge system stops the supply of chlorine to the entire unit, including the evaporator and the superheater (Fig. 2.48). Once repairs have been carried out, chlorine flow to the plastics production unit can be restarted.

The chlorine flow is 0 kg/h during repairs. While the difference between the required and actual flows to the production unit remains very high (Fig. 2.49), the alert remains active. This dual safety measure provides functional redundancy for

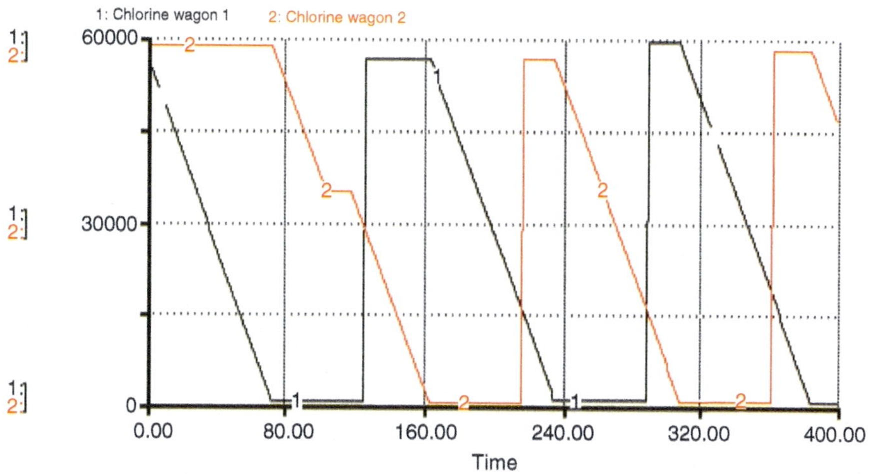

Fig. 2.45 Temporary shutdown of the discharge of chlorine wagon 2 following a small chlorine leak at $t = 100$ h. The system will be temporarily shut down for 15 h in order to repair the part. After this time, normal discharge activities resume

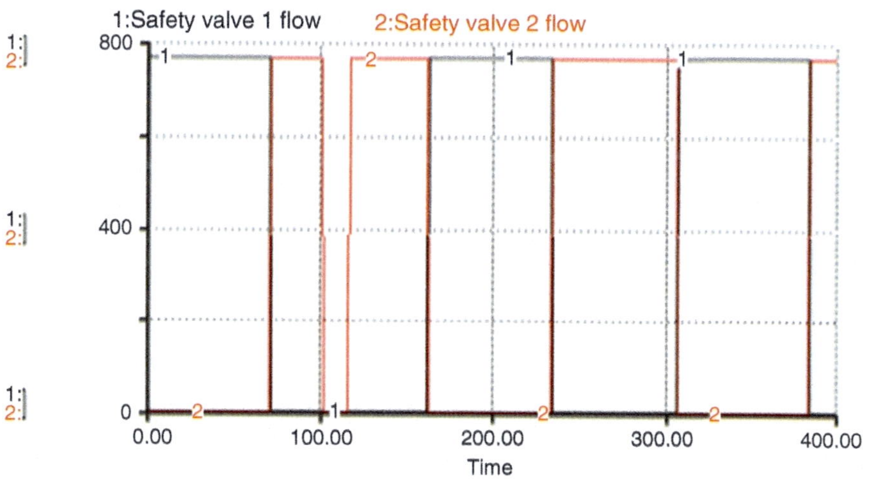

Fig. 2.46 Chlorine flow passing through safety valves 1 and 2. If the sensor detects a chlorine leak during the discharge of wagon 2, safety valve 2 is closed

the shutdown of the discharge unit and the transportation of chlorine. The functional redundancy principle is an important element in the formalization of defence-in-depth measures (Garbolino et al. 2005).

These examples show that it is possible to refine the model to simulate abnormal scenarios and the effects of prevention measures. Such an approach has many advantages, but there are also some limitations that are addressed in the following section.

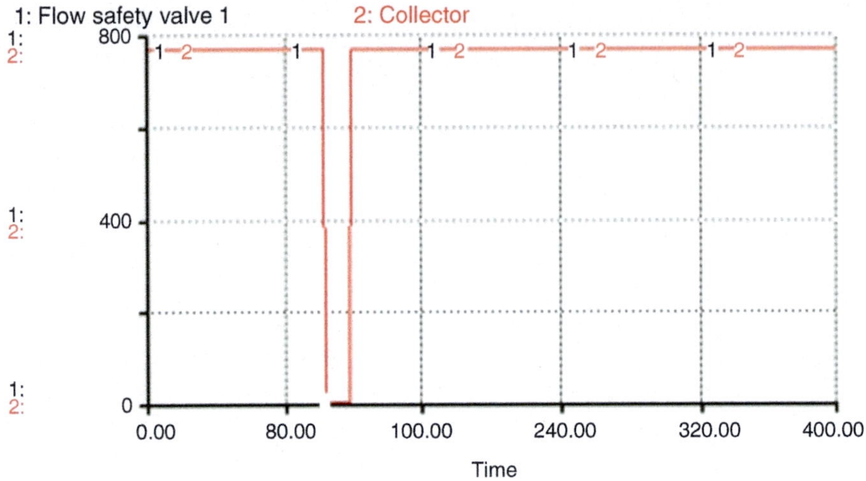

Fig. 2.47 Chlorine flow passing through the collector and control valve 1

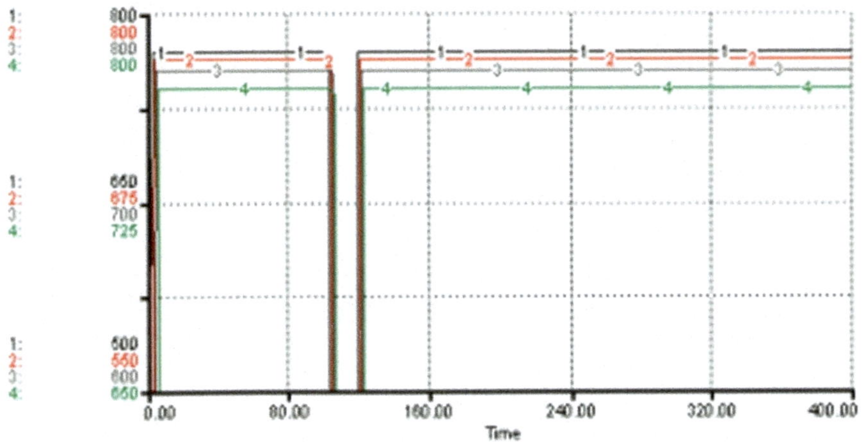

Fig. 2.48 Chlorine flow in the evaporator and superheater: this flow stops when the leak sensor is activated

2.4 Evaluation of Benefits and Limits

The approach presented here combines methods and tools that can be used in a dynamic risk analysis of an industrial system, such as a unit for the discharge and transfer of chlorine. This approach is intended to be generic, and so may be of interest to both industry and government departments that are responsible for monitoring industrial facilities.

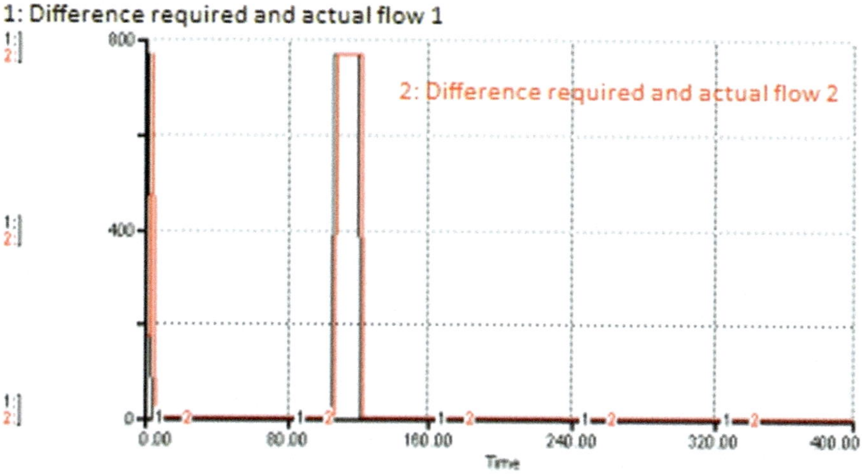

Fig. 2.49 Difference between the required flow and the actual flow at the outlet of the evaporator (1) and the superheater (2)

The methods and tools used here consist of:

- A dynamic approach to modelling the system and the simulation of its behaviour using the STELLA® software package;
- Risk analysis using the HAZOPS method;
- The simulation of an atmospheric release using the PHAST® software package to simulate failure scenarios identified in the risk analysis;
- A second dynamic model of the system that simulates system behaviour once prevention and protection measures have been implemented.

There are advantages and disadvantages to this approach. The main advantages are:

- *A dynamic approach to risk management:* the key contribution is the ability to simulate systems in a range of failure scenarios. This makes it possible to run real-time simulations to explore the consequences of these failures for the workforce, for local residents and for the environment. When coupled with a tool that can estimate the toxic effects of atmospheric emissions, dynamic modelling can be used to investigate the kinetics of events, the amount of hazardous material released over a certain period of time, the response times of protection measures and operators, etc. The whole system—technical, human and organizational aspects—is modelled, together with any prevention and protection measures that may be introduced. In other words, this modelling approach provides support for the development of a systemic, integrated and dynamic approach to risk management.
- *A continuous process of improved risk management:* the use of a dynamic system modelling platform (such as STELLA®) makes it possible to represent and simulate the behaviour of industrial systems. It is also very easy to modify the model

and re-run it to check its relevance. These modifications can be introduced at various points, such as: during the system design phase; when the hazard study is updated; when technical or organizational changes are introduced; and when safety plans are formalized. The dynamic modelling approach can therefore be included as part of a continuous improvement strategy to manage industrial risks.

- *A way to validate the ISE*: the identification of accident scenarios makes it necessary to identify prevention and protection measures, including ISE. By including these in the model, their effectiveness during the unfolding of accidents can be investigated. Dynamic modelling therefore raises a question about who is the right person to evaluate the relevance of any new safety measures.
- *A better representation of the system:* Implementing the model means that the developer—usually a process expert—has to identify the system elements, define and set variables, and run simulations to verify the model. These tasks provide the modeller with different types of knowledge about the industrial system that complement the standard process engineering perspective. The whole approach requires the user to think about the function and behaviour of the system in ways that are not typically part of process engineering expertise.
- *Support for communication and training:* System dynamics uses simple representations, and the results of modelling can be shown graphically by software packages such as STELLA®. It becomes relatively easy for the expert modeller to explain the system's behaviour and the role of safety measures to others (the workforce, consultants or the official inspection bodies) during the failure scenarios defined in the hazard study. This approach can also be used in a participatory manner, by setting up expert groups to conduct or evaluate hazard studies. A dynamic modelling platform can also be used to support workforce training by simulating each stakeholder's activities.

The main disadvantages are:

- *The level of complexity*. As the level of complexity in the model rises, so does the level of detail involved. This means that there will be an increase in the number and nature of the variables that have to be defined, and the ways that they interact in causal loops.
- *Difficulties in interpreting the results*. As the model and simulation become more complex, it becomes more difficult to interpret the results and how they relate to the system's behaviour. The length of time that must be spent on this aspect of modelling can vary significantly depending on the degree of complexity represented in the model.

These disadvantages can be overcome by finding an appropriate balance between the time spent on modelling the system and the time spent on the dynamic risk analysis. This balance also depends on the experience of the modeller, both in terms of modelling methods and their knowledge of industrial processes.

The next chapter describes an application of a systemic model applied to assess the risk in a chemical industry, by taking into account human, technical and organizational dimensions.

References

Andrews JD, Moss TR (2002) Reliability and risk assessment. Wiley, Chichester

Aracil J (1984) Introduction à la dynamique des systèmes. PUL, Collection Science des systèmes, 412 p

Deaton ML, Winebrake JJ (2000) Dynamic modeling of environnemental systems, coll, Modeling dynamic systems, Springer, New York, 194 p

Donnadieu G, Karsky M (2002) La systémique, penser et agir dans la complexité. Editions Liaisons, Paris, 269 p

EuroChlor (2005) Guideline for site security of chlorine production facilities. GEST 05/316, 11 p

EuroChlor (2008) The chlorine reference manual. GEST 06/317, 62 p

Forrester JW (1961) Industrial dynamics. MIT Press, Cambridge, p 464

Garbolino E, Guarnieri F, Cambon J (2005) Can the nuclear safety concept "Defense In Depth" improve the security of industrial and technological activities in France? Chem Eng Trans 6:533–538

Garbolino E, Chery JP, Guarnieri F (2009) Dynamic systems modelling to improve risk analysis in the context of seveso industries. Chem Eng Trans 17:373–378

INERIS (2004) Application de la classification des substances et préparations dangereuses à la nomenclature des installations classées, 29 p

INERIS (2008) Émissions accidentelles de substances chimiques dangereuses dans l'atmosphère. Seuils de toxicité aigüe du Chlore. DRC-08-94398-10645A, 3 p

Kjellen U (2000) Prevention of accidents through experience feedback. CRC Press, London, 424 p

MEDAD (2007) Nomenclature des Installations Classées. Liste des activités soumises à la TGAP. Direction de la Prévention des Pollutions et des Risques. Service de l'Environnement Industriel, 76 p

MEDD, DPPR, SEI, BARPI (2004) Accidentologie du chlore, 65 p

NIOSH (2007) NIOSH pocket guide to chemical hazards. National Institute for Occupational Safety and Health Education and Information Division, 424 p

Chapter 3
System Dynamics Applied to the Human, Technical and Organizational Factors of Industrial Safety

Hafida Bouloiz and Emmanuel Garbolino

In this chapter, system dynamics is applied to a socio-technical system in the chemical industry. The objective is to develop a dynamic model of the industrial system, which formalizes the causal interdependencies between the factors that define the safety conditions for that system. A systemic approach is applied to the case study of an chemical storage facility. The implementation, which is supported by Vensim® software, makes it possible to include the human, technical and organizational factors that all contribute to safety management (Bouloiz et al. 2013).

This chapter begins with a brief overview of the Vensim® software package for system dynamics modelling. Then the case study is described. The development of the model of the facility's safety conditions is detailed and its implementation in the Vensim® interface is presented. Finally, several case study simulations that generate dynamic system behaviour are described.

3.1 Vensim® and System Dynamics Modelling

Vensim® is a software package that supports system dynamics modelling. It is based on Forrester's (1961) stock–flow approach. The software translates the relationships between the system's structural elements, represented diagrammatically, into a set of equations. Models in Vensim® are constructed using the same principles as STELLA, which was described in Chap. 2. The Vensim® interface has a toolbar

H. Bouloiz
Ecole Nationale des Sciences Appliquées, Agadir, Morocco

E. Garbolino (✉)
MINES ParisTech/PSL Research University, CRC, Sophia Antipolis Cedex, France
e-mail: emmanuel.garbolino@mines-paristech.fr

F. Guarnieri, E. Garbolino (eds.), *Safety Dynamics*, Advanced Sciences and Technologies for Security Applications,
https://doi.org/10.1007/978-3-319-96259-7_3

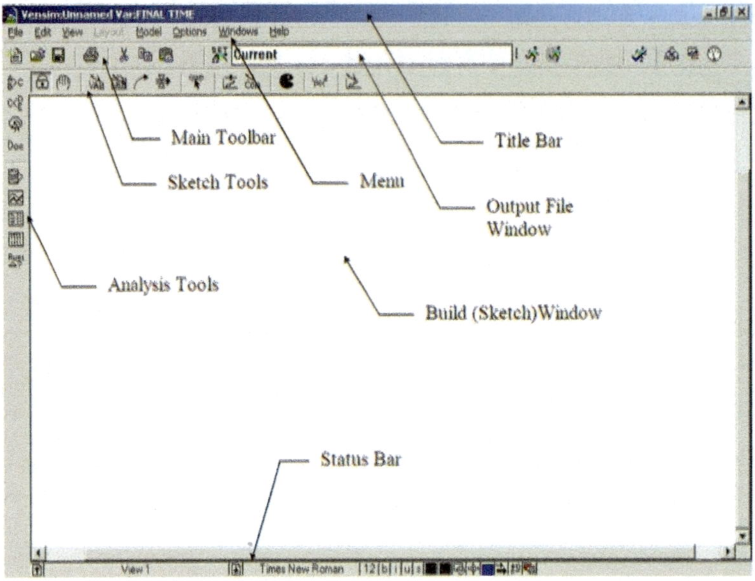

Fig. 3.1 The Vensim® interface. (Ventana systems)

containing a set of icons for the construction of models of system dynamics (Fig. 3.1). Its key feature is the ability to perform graphical analysis of causal trees.

More detailed information about the features of Vensim® and instructions for its operation can be found in the Vantana System Reference Manual (1999).

The industrial system that was used in the case study is described in the following section. The goal was to use system dynamics modelling to formalize the interactions between the human, technical and organizational factors in this system.

3.2 The Industrial System

The system is a storage facility that is used in the manufacturing of chemicals for industrial use. The site is owned by an industrial group that is a world leader in the field of specialty chemicals. The group is organized into several operating divisions: chemicals for textiles, leather and paper (e.g., polymer dispersions, sulphur dyes, acids, silicones); pigments and additives (e.g., polymeric additives, flame retardants, waxes); and products with chemical functions (e.g., detergents, products for human use). The site consists of the following units (as shown in Fig. 3.2):

- a product storage unit;
- a manufacturing unit;
- a maintenance unit; and
- an analysis laboratory.

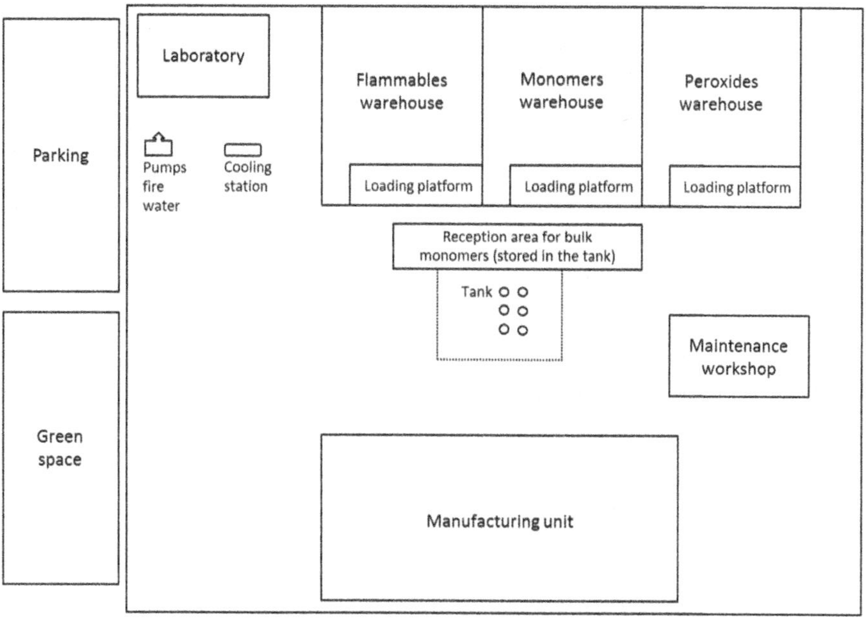

Fig. 3.2 Configuration of the industrial site

Bulk chemicals are stored in tanks; non-bulk chemicals are stored in warehouses marked with signs indicating the type of chemicals they contain. Incoming bulk materials are handled using forklift trucks and pallet trucks. Forklift trucks stack products on shelves and load and unload delivery trucks; pallet trucks transport products between on-site locations.

The industrial facility includes three warehouses, one for each type of products: monomers (acrylic acid) in liquid state; peroxides in solid state; and flammable products in liquid state. Monomers have a risk of polymerization, which is a highly exothermic process that can cause an explosion. Peroxides are characterized by their oxidizing and combustive properties. They can foster or activate the combustion of a flammable substance. Flammable products ignite when exposed to air and then continue to burn.

The chemical containers must be stored at particular temperatures based on the type of chemical that they contain. Monomers must be stored in the range 16–25 °C to prevent their polymerization or solidification. Peroxides must not be stored at above 30 °C, and flammable products should not be stored at above 35 °C.

The case study focuses on the warehouse for flammable products. This storage unit is represented as a complex system of interacting technical, human and organizational elements, which need to be properly adjusted and organized to minimise risks. Each operator has responsibilities that contribute to the smooth running of the facility. The facility's operating procedures are a very important organizational element. These procedures are outlined in comprehensive documentation that includes:

storage instructions; safety data sheets; safety measures for use in the event of the accidental spillage of a product; a list of safety checks; and location cards. Technical tools include safety devices in the forms of preventive equipment (alarms, heat detectors) and protective equipment (cooling systems, personal protection equipment, sprinklers).

An accident can be caused by an organizational malfunction (inadequate procedures, a failure to update instructions, lack of feedback), by human error (inappropriate behaviour, a failure to comply with procedures, lack of monitoring) and by technical failure (safety devices). Other human-related factors that should be taken into account include motivation, stress, expertise, training, and the working environment.

All of the variables and parameters that are considered here are specific to the industrial facility. However, other variables and parameters could be added, which would extend the range of safety conditions that could be considered. A more detailed analysis of the safety behaviour of operators, and a summary of the main factors that influence them, is presented in Chap. 4.

The next section describes the system dynamics model used in this case study. The model formalizes the causal interactions between the elements that define the safety conditions at the level of the industrial system.

3.3 Modelling and Dynamic Analysis

The dynamic model was developed using Forrester's (1961) modelling process, adapted to this study (Bouloiz et al. 2013), based on the following four steps:

1. Problem articulation: defines the purpose of the model and identifies the entities and interactions to highlight.
2. Causal diagram development: qualitatively assesses the relationships and causal interactions between the different system elements.
3. Stock–flow diagram development: introduces the temporal dimension into the model by considering the rate of change or evolution of the levels of variables (stock and flow) over time.
4. Simulation modelling: allows an analysis of the dynamic behaviour of the system.

The remainder of this chapter explains these steps in detail.

3.3.1 Problem Articulation

Risk management is based on the control of each system element and their interactions. In this case, temperature maintenance and stock control are the two main safety conditions. The safety conditions are defined by the model's variables, which are represented by safety measures. Specifically, system dynamics was applied to formalize the causal interdependencies between the human, technical and organizational factors that define the safety conditions at the flammable materials storage facility, and to highlight the system's emergent behaviours.

3.3.2 *Causal Diagram*

This step involves linking the variables that were identified during problem articulation using Forrester's (1961) causal diagram principles. The definition of the interactions between the behavioural factors considered in this application was based on previous studies (Table 3.1).

Table 3.1 Explanation of interactions between selected behavioral factors

Behavioral factors	Previous studies	Summary	Definition of the interactions
Motivation	Karsky and Donnadieu (1990), Ryan and Deci (2000), Jones (2005) and Kaplan and Norton (1992)	Motivation directly influences the decision to act of the worker and therefore his behaviour	
		Motivation positively influences the behaviour	
		The sources of motivation are present in the organizational work environment. A favorable work environment increases the workers's motivation	
Stress	Giambiasi et al. (2005), Frere, (2000), Jones, (2005), Marchand et al. (2005) and Hart and Cooper (2001)	Stress is one of the factors studied to analyze the behavior of the human factor in the workplace	
		Stress has a negative impact on the behaviour	
		Stressors are the sources of stress in the work environment. A favorable environment reduces stress of workers	
		Occupational stress reduces the motivation of workers	
Competence	Denis-Remis (2007), Boucher and Burlat (2003), Bonjour and Dulmet (2002) and Cesar 2005	The training is an efficient tool to implement safety behaviour	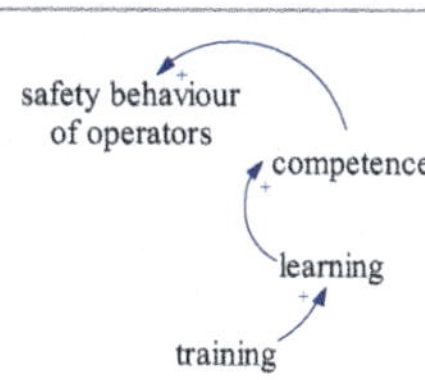
		Learning through training is an important source of skills development	

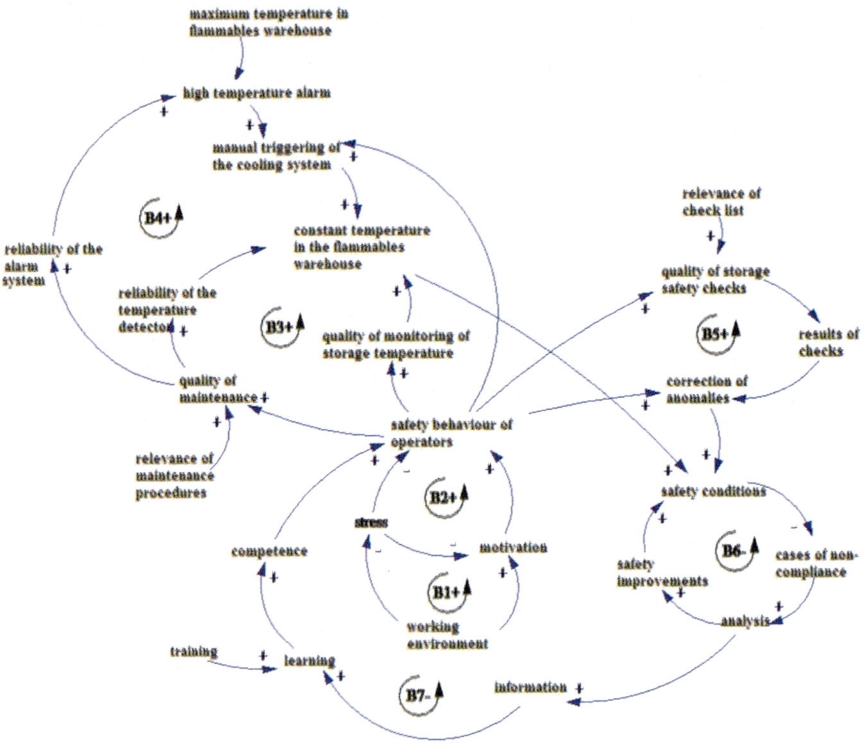

Fig. 3.3 Causal diagram of safety conditions at the flammable materials warehouse

Figure 3.3 shows the causal diagram for the system that was developed using Vensim® software. This causal diagram represents a set of feedback loops and formalizes the causal interdependence between the system's safety conditions.

Loop B1 represents the condition where a good working environment positively influences (increases) operators' motivation, and negatively influences (reduces) their stress levels. Loop B2 shows that the operators' motivation positively influences (improves) their safety behaviour.

The relevance of maintenance procedures, together with operators' behaviour, influences the quality of maintenance that, in turn, influences the reliability of safety devices (the alarm system and temperature sensor) installed in the warehouse. Maintaining the correct temperature (loop B3) is a function of: how well operators monitor the temperature; the reliability of the alarm system; the reliability of the temperature detector (sensors); and the cooling system that is manually activated when the warehouse temperature exceeds 35 °C. The activation of this cooling system is a function of operators' safety behaviour because they have to manually intervene when the temperature alarm is triggered (loop B4). The quality of storage safety checks and the correction of anomalies (identified by these checks) are a function of operators' behaviour (loop B5).

Non-compliance can occur with respect to product storage checks or maintenance of the storage temperature. In this case, the situation must be analysed from two perspectives. First, to improve safety (loop B6), and second to improve operators' behaviour: the analysis provides a source of information for learning and acquiring additional knowledge (loop B7). Learning through training positively influences both the operators' competence and their behaviour.

3.3.3 *Stock–Flow Diagram*

The stock–flow diagram corresponding to the causal diagram described above is shown in Fig. 3.4. This demonstrates the relationships between the model's variables. This model can be used to carry out sensitivity studies using changes in control variables, for example. The definitions of the variables used to model safety conditions at the site are shown in Table 3.2.

In system dynamics, all simulations have a specified initial state. Before running the simulation, it is necessary to define the initial conditions for each state and stock variable. The initial conditions used in this application came from the company's own audit data. This audit targeted particular points to be controlled at all industrial units.

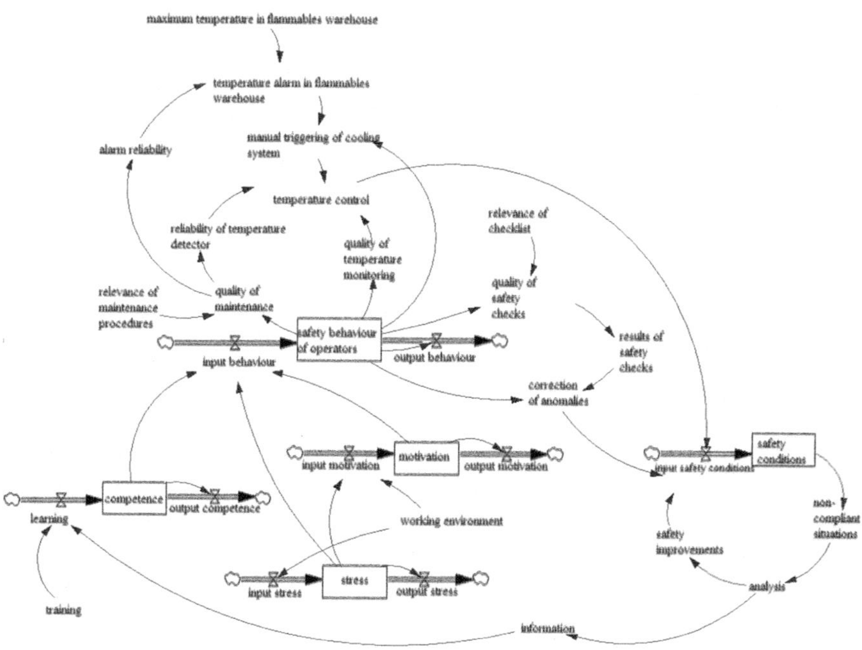

Fig. 3.4 Stock–flow diagram of safety conditions at the flammable products warehouse developed using Vensim® software

Table 3.2 Variables used to model safety conditions at the industrial facility

Variable	Definition	Function
Temperature alarm in the warehouse	IF THEN ELSE (warehouse temperature > 35, 1, alarm reliability)	Activated when the temperature in the warehouse exceeds 35 °C
Temperature control in the warehouse	Reliability of temperature sensor + quality of monitoring of storage temperature + triggering of cooling system	Temperature control is a safety condition that avoids the risk of fire or explosion
Quality of maintenance	Safety behaviour of operators + relevance of maintenance procedures	Quality of maintenance influences the functioning and reliability of equipment
Quality of temperature monitoring	A function of operator behaviour	Monitoring of storage temperature is a safety condition
Quality of safety checks.	A function of operator behaviour and the relevance of checklists	Safety checks can identify anomalies (e.g., packaging, shelving)
Behaviour of operators	State variable	Operator behaviour
	INTEG (behaviour input – behaviour output)	
Stress	State variable	Level of stress
	INTEG (stress input – stress output)	
Motivation	State variable	Level of motivation
	INTEG (motivation input – motivation output)	
Safety conditions	State variable	Safety conditions in the warehouse
	INTEG (safety conditions input, 0.5)	
Non-compliance	1/safety conditions	A function of safety conditions
Safety improvement	A function of non-compliant situations	Safety improvement is an important condition
Correction of anomalies	A function of the results of safety checks	Anomalies identified during safety checks are corrected
Information	A function of non-compliant situations	Analysis of non-compliance is a source of information that improves the competence of operators
Learning	Flow	Learning is a function of training and knowledge acquisition from the analysis of non-compliant situations
Competence	State variable	Level of competence
	INTEG (learning – Competence output)	
Working environment	0–1	Working environment influences the behaviour of operators
	Poor-quality working environment = 0	
	High-quality working environment = 1	

(continued)

Table 3.2 (continued)

Variable	Definition	Function
Training	0–1	Training is a source of operator competence
	Low level of training = 0	
	High level of training = 1	
Relevance of checklists	0–1	Checklists are used in safety checks of warehouse storage
	Irrelevant checklist = 0	
	Highly relevant checklist = 1	
Relevance of maintenance procedures	0–1	Procedures are used during maintenance of safety devices
	Irrelevant procedures = 0	
	Highly relevant procedures = 1	

Table 3.3 Case study conditions

Case study	Data	Description
Case 1: Training	*Current situation*	0.75
	High level	Increase of 20%
	Low level	Decrease of 20%
Case2: Quality of working environment	*Current situation*	0.75
	High quality	Increase of 20%
	Low quality	Decrease of 20%
Simulation period (days)		Start time: 0
		End time: 6
Time step		0.25 days

The audit's results were used to set initial values for the variables in the model. An evaluation grid based on a qualitative scale (very low; low; medium; high; very high) was developed and the model's variables were weighted according to these results and then mapped onto a linear numerical scale covering the range 0 (very low) to 1 (very high). This operation was carried out in cooperation with the company and subsequently validated by them.

3.3.4 Simulations and Results

The initial analysis explored two aspects: the impact of the variable *training* and of the variable *working environment* on safety conditions. In each case, two scenarios were presented: a 20% of the initial value increase and a 20% decrease in both variables. The 20% range was chosen as representative of a potential evolution in safety conditions at the facility.

The following table shows the simulation conditions in each case study (Table 3.3).

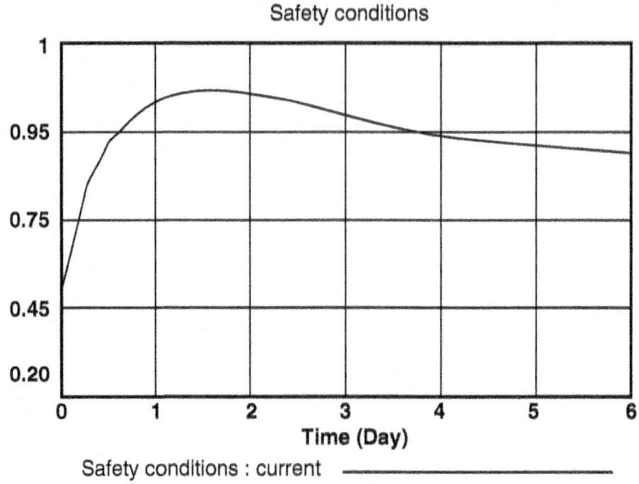

Fig. 3.5 Simulation of the variable *safety conditions* in the normal (current) situation

The values of the variable *current situation* (of training) are estimates taken from the original safety audit. The current situation is seen here as referring to normal operating conditions, before any changes are introduced. The simulation period is arbitrarily set to 6 days.

In general, in this system, the prevention and reduction of at-risk situations provides safe working conditions. The level of these conditions is shown in Fig. 3.5, using the dataset from the current situation. This figure shows the state of the variable *safety conditions* in the current situation, i.e., the normal situation without any modifications to the model's variables.

The effects of *training* on safety conditions were simulated (Fig. 3.6). The graph shows that an increase in the level of training has a small effect on safety (dashed and dotted line) compared to the normal situation (solid line). On the other hand, a lack of training (i.e., poorly trained and therefore less-skilled operators) leads to a decrease in the level of safety conditions (dotted line). These results indicate that managers must at least maintain the current level of training; increasing it only appears to have a small effect.

The effects of the quality of the *working environment* on *safety conditions* are shown in Fig. 3.7. The graph shows that a deterioration in the working environment (a poorer-quality working environment) has a large effect on safety (dotted line), while a more favourable environment improves the system's safety conditions (dashed and dotted line).

These results can be explained by examining the direct impact of the working environment on the operators' behaviour as shown in Fig. 3.8. There is a large effect of the quality of the working environment on operators' behaviour.

As the causal diagram previously shown in Fig. 5.9 highlighted (in particular feedback loops B1 and B2), the variable *working environment* influences the two behavioural factors *motivation* and *stress*. When the working environment is favour-

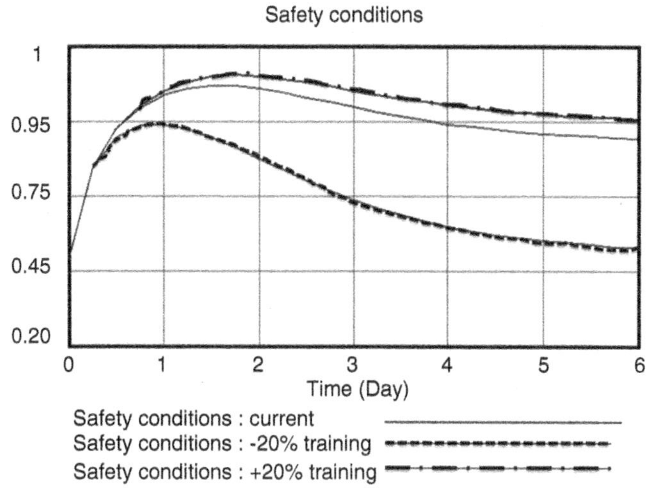

Fig. 3.6 Effects of *training* on the variable *safety conditions*

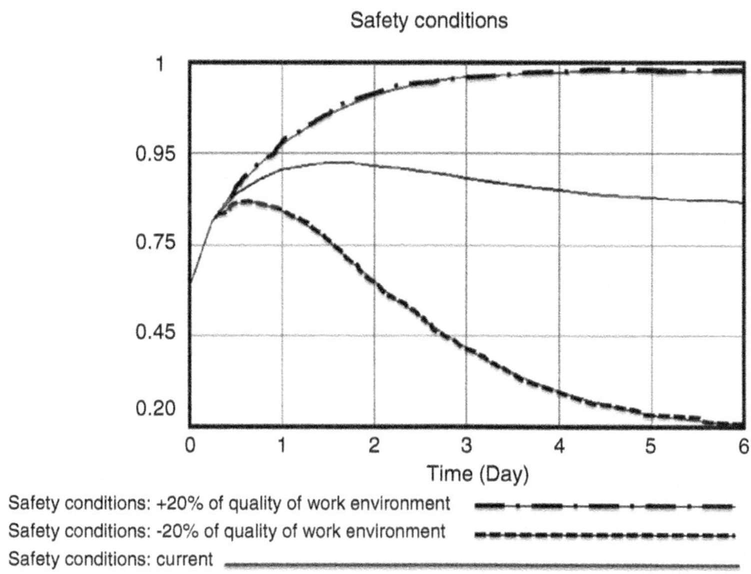

Fig. 3.7 Effects of the variable *quality of the working environment* on *safety conditions*

able, motivation increases and stress decreases, which positively affects operators' behaviour. The results of these simulations suggest that the company should maintain a high level of operator training, and implement a continuous improvement process with respect to the quality of the working environment.

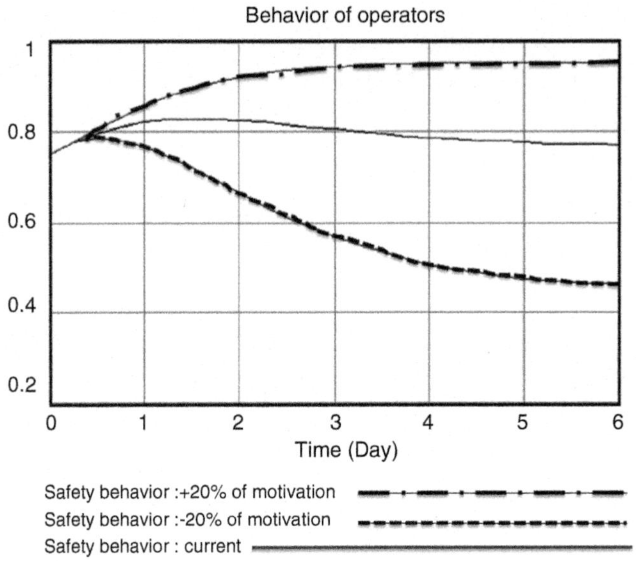

Fig. 3.8 Effects of the variable *working environment* on *operator behaviour*

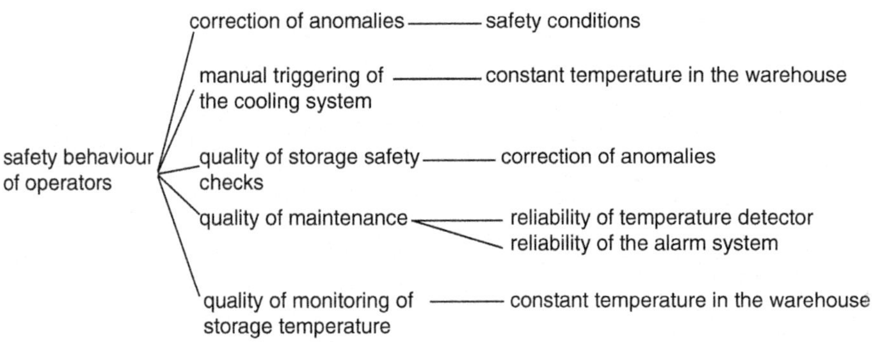

Fig. 3.9 Use tree of the variable *operator behaviour*

Operators' behaviour influences the quality and success of all safety actions at the facility. The tree shown in Fig. 3.9 illustrates the relationship between the operators' behaviour and the risk management factors that define safety conditions at the facility.

This tree highlights the importance of the safety behaviour of operators as a major challenge in the overall safety of the system. Operators carry out the following activities:

- Correction of abnormal situations to avoid a potential accident;
- Monitoring of the storage temperature to maintain the correct system temperature;

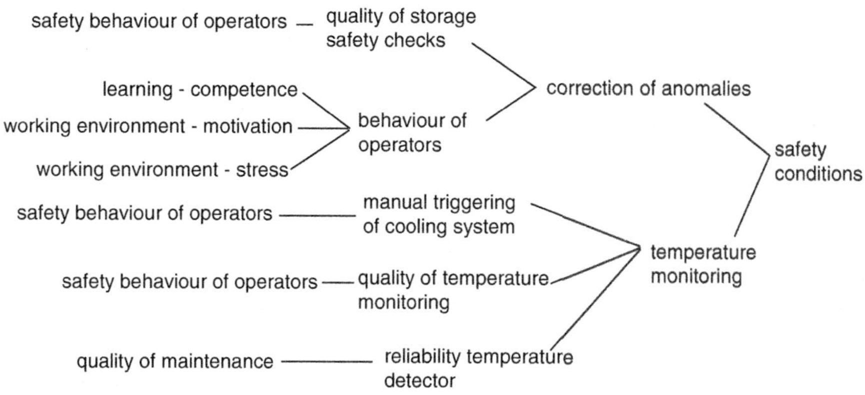

Fig. 3.10 Causal tree of the variable *safety conditions*

- Maintenance of equipment that influences the reliability of the safety devices;
- Checks that lead to the correction of any anomalies;

All of these actions that define the safety conditions at the facility are shown in the causal tree in Fig. 3.10. This tree provides a complete description of the various feedback loops that are shown in Fig. 3.4, and highlights the causal interdependencies between the various safety conditions.

Taken together, the results of modelling this application are encouraging. They provide a useful demonstration of how modelling can be used to improve understanding of the safety conditions of a system, and contribute to ongoing improvements to system safety.

References

Bonjour E, Dulmet M (2002) Articulation entre pilotage des systèmes de compétences et gestion des connaissances. 1er colloque de gestion des compétences et des connaissances en génie industriel, Nantes, 12–13 décembre, pp 43–50

Boucher X, Burlat P (2003) Vers l'intégration des compétences dans le pilotage des performances de l'entreprise. JESA 37(3):363–390

Bouloiz H, Garbolino E, Tkiouat M, Guarnieri F (2013) A system dynamics model for behavioral analysis of safety conditions in a chemical storage unit. Saf Sci 58:32–40

Cesar M (2005) Le processus d'apprentissage organisationnel: vers de nouveaux modes d'apprentissage et de gestion des compétences. Actes du 7ème Colloque International de Psychologie Sociale, ADRIPS, Rennes

Denis-Remis C (2007) Approche de la maitrise des risques par la formation des acteurs. Thèse de doctorat. Ecole des Mines de Paris

Forrester JW (1961) Industrial dynamics. MIT Press, Cambridge, p 464

Frere R (2000) Contribution à l'intégration d'aspects humains dans la modélisation des systèmes de fabrication : vers une gestion conjointe des ressources humaines et de la production. Thèse de doctorat, Université de Valenciennes et du Hainaut-Cambrésis, Valenciennes, France

Giambiasi N, Frydman C, Seck M (2005) Modélisation et simulation du comportement humain avec le formalisme devs. In Actes de 6ème Conférence Francophone de Modélisation et Simulation (MOSIM'06), Maroc
Hart P, Cooper C (2001) Occupational stress: toward a more integrated framework, vol 2, chapitre Handbook of industrial work and organizational psychology. Sage, London
Jones C (2005) Behavioral theory in simulation: ambiguous results from simple relationships. In: Proceedings the 23rd international conference of the System Dynamics Society, USA
Kaplan RS, Norton DP (1992) The balanced scorecard – measures that drive performance. Harv Bus Rev 70(1):71–79
Karsky M, Donnadieu G (1990) The dynamic of behavior and motivation. In: Proceeding of the international system conference, Boston
Marchand A, Demers A, Durand P (2005) Does work really cause distress? The contribution of occupational structure and work organization to the experience of psychological distress. Soc Sci Med 60:1–14
Ryan RM, Deci EL (2000) Intrinsic and extrinsic motivations: classic definitions and new directions. Contemp Educ Psychol 25:54–67
VENSIM (1999) Reference manual version 4, Ventana System, Inc

Chapter 4
Modelling and Dynamic Analysis of Safety Behaviour

Hafida Bouloiz and Emmanuel Garbolino

This chapter presents a dynamic analysis of the safety behaviour of operators and the organizational environment in which they work. This environment contains a set of factors (variables) that affect human behaviour in the context of industrial safety. These behavioural factors are identified from an initial analysis of the principal factors that influence how operators behave safely. The focus of this chapter is the explanatory factors that underlie safe behaviour and the causal relationships that link these factors, from a combined perspective of system dynamics and fuzzy logic. System dynamics is used to accurately define the causal interactions between this set of explanatory factors and represent them from a systemic and dynamic viewpoint. Fuzzy logic is subsequently used to take into account the qualitative and inexact nature of the variables involved. Taken together, system dynamics and fuzzy logic is used to develop a model that more accurately describes the safe behaviour of operators.

4.1 The Value of Human and Organizational Factors in Industrial Safety

The complexity of industrial systems poses a challenge for safety, because it can be a source of deviations in system behaviour. These deviations are themselves among the attributed causes of incidents and accidents. System complexity is related to the unpredictable (and emergent) aspects of the system's behaviour that arise from the

H. Bouloiz
Ecole Nationale des Sciences Appliquées, Agadir, Morocco

E. Garbolino (✉)
MINES ParisTech/PSL Research University, CRC, Sophia Antipolis Cedex, France
e-mail: emmanuel.garbolino@mines-paristech.fr

F. Guarnieri, E. Garbolino (eds.), *Safety Dynamics*, Advanced Sciences and Technologies for Security Applications,
https://doi.org/10.1007/978-3-319-96259-7_4

interactions between its many elements (technical, human, organizational and environmental).

Industrial systems tend to be susceptible to accidents that are caused by failures in the interactions between the components, rather than the simple failure of a single component (Leveson 2004), as a result of the manifold relationships between the various system elements. Thus, for example, the unexpected failure of a component may be due to poor maintenance (an organizational factor) or the use of a poorly designed component (a technical factor). Accidents cannot simply be attributed to human actions, and labelled as *human error* (the human factor). When an accident occurs, the person may have followed a particular course of action because it appeared to be the right thing to do in that situation, or because the way that the system was designed meant that it was the only action that was open to them.

The inclusion of the human factor and its organizational environment in industrial risk management is regarded here as being central to improvements in industrial safety. This environment, however, contains many factors that affect human behaviour. When analysing such situations, the complex and diverse nature of the problem becomes clear, given the interplay between the different types of factors involved: psychological, psychosocial, organizational and economic. Furthermore, the complexity of human behaviour means that it cannot be quantitatively assessed with precision and certainty.

In order to address the difficulties, a methodology was developed which combined two techniques:

- System dynamics (described in Chap. 1)
- Fuzzy set theory, which is used to account for the qualitative, uncertain and linguistic nature of the variables that influence the safety behaviour of the human factor.

4.2 The Principal Criteria of Safety Behaviour in the Human Factor

There are two main schools of thought about the concept of *behaviour*. Behaviourist psychologists such as Watson (cited in Bloch 1994 p. 153), take behaviour to be "the set of objectively observable reactions that an organism that usually has a nervous system executes in response to environmental stimuli, which are themselves objectively observable." In contrast, cognitivist approaches take a broader view and define behaviour as "*a comprehensible reality in the form of units of observation, acts, the frequency and sequences of which are likely to be modified; it translates into action the image of the situation as it develops, with its own tools, through the entity that is being studied: behaviour expresses a form of representation and construction of a particular world.*" (e.g., Gallo 1999).

Similarly, two types of safety behaviour can be distinguished (Simard and Marchand 1997). The first corresponds to prudence, which consists, for example, of respecting and following safety rules. An example is wearing protective equipment. The second relates to safety (behaviour) initiatives. These initiatives include activities such as suggestions from operators about how to avoid a risk or improve safety, and requests from operators for corrective action when equipment is found to be defective.

Human behaviour in the workplace has been widely studied, both from the perspective of a single aspect, such as competence (Harzallah and Vernadat 2002), personality (McCrae and John 1992; Howard 2006), performance (Siebers et al. 2004), and from the perspective of the psychological and psychosocial factors that influence behaviour, such as stress (Giambiasi et al. 2005; Frere 2000), motivation (Karsky and Donnadieu 1990), fatigue (Giambiasi et al. 2005; Karsky and Adamo 1977), the need for accuracy (Belyavin and Farmer 2006), self-efficacy (Bandura 1997), and conflict (Cornet 1995; Frere 2000).

The causal relations between the psychological factors that influence worker behaviour have mostly been studied in a pairwise fashion. Jones (2005), for example, explored the impact of variation in effort on the performance and the effect of motivation on effort (Jones 2005).

Elkosantini (2007) incorporated the causal interactions between psychological factors (motivation, stress, performance, satisfaction, self-efficacy) and psychosocial factors (conflict, communication) a model of operator behaviour. This model was used to study the effects of these two types of factors on the efficiency of operators in a production system.

There are other organizational factors that affect operator behaviour too. Simard and Marchand (1997), for example, show that micro-organizational factors such as the degree of internal group cohesion, or the extent of cooperation between the group and the supervisor can determine the propensity of operators to act prudently and take safety initiatives. They also show that macro-organizational factors, such as the degree of commitment of senior executives to prevention, are important and positively influence the safety behaviour of the workforce.

The following section presents a systemic modelling methodology that combines system dynamics and fuzzy logic to provide a generic model of safety behaviour of the human factor.

4.3 Proposed Generic Model of the Safety Behaviour of Operators

4.3.1 Principles

Several psychological, psychosocial, organizational and economic factors have to be taken into account when developing a representative model of safety behaviour. The aim is not to exhaustively study all of the factors that influence human behaviour, but to develop a representative model of the context for human action in industry that is both detailed and complex. The factors that have been selected for the analysis of human behaviour (primarily in the workplace) have been the subject of many scientific studies (Harzallah and Vernadat 2002; McCrae and John 1992; Howard 2006; Siebers et al. 2004; Giambiasi et al. 2005; Frere 2000; Karsky and Donnadieu 1990; Belyavin and Farmer 2006; Jones 2005; Elkosantini 2007).

4.3.1.1 Modelling Methods

The proposed modelling methodology combines system dynamics and fuzzy set theory. System dynamics was previously described in Chap. 1, and illustrated in Chaps. 2 and 3.

4.3.1.2 Fuzzy Set Theory

The theoretical foundations of fuzzy logic were established by Zadeh (1965). The technique is an extension of Boolean logic based on the mathematical theory of fuzzy sets, which is a generalization of the classical set theory (Dernoncourt 2011). The approach is based on human reasoning rather than complex calculations.

Fuzzy set theory is the formalism that is best-suited to the qualitative description of linguistic variables (Zadeh 1965, 1975, 1978). It provides a formal framework for modelling the imprecise and uncertain aspects of natural language, and allows designers to better understand natural phenomena that are imprecise and difficult to model. The approach is based on the definition of rules and membership functions applied to fuzzy sets (Dernoncourt 2011). The value of the approach here lies in the formalization of linguistic variables that can be used to model imprecise concepts such as a *low stress level*, or a *high level of motivation*. Fuzzy sets can be used to represent these variables linguistically.

A linguistic variable is a triple (V, X, T_V), where V is a variable that is defined on a reference set X. The finite or infinite set $T_V = \{A1, A2, \ldots\}$ contains fuzzy subsets normalised from X, which can be used to characterize V. The number of members in the set T_V depends on the goal of the modelling. Figure 4.1 presents an example of the representation of the linguistic variable *motivation*. In this example

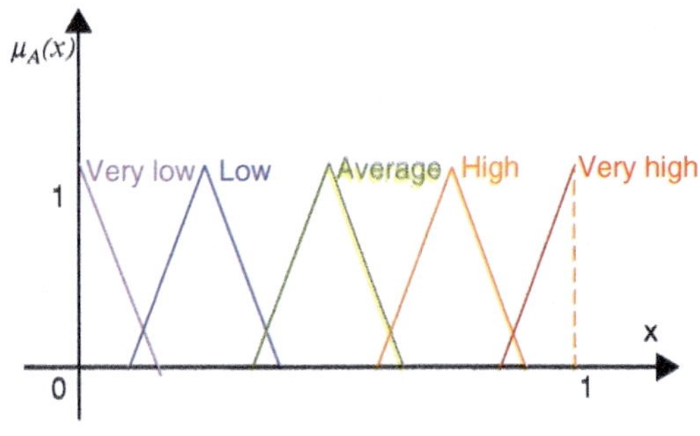

Fig. 4.1 The linguistic variable *motivation*

V = *motivation*; *X* = *R* (the set of real numbers); and *Tv* = *{very low, low, average, high, very high}*.

Fuzzy reasoning, which is also called *approximate* reasoning, is based on fuzzy rules. These are expressed in natural language using linguistic variables (Dernoncourt 2011; Bezdek and Sanker 1992; Novák and Perfilieva 1999). Rules can be expressed in the general form:

If condition 1 then Action 1 or
If condition 2 then Action 2 or
If....

The conditions may depend on several variables interconnected by operators OR/AND. For example, here some rules of conduct that a driver follows (Dernoncourt 2011):

If the light is red if my speed is high and if the fire is close then I brake hard.
If the light is red if my speed is low and if the fire is far so I maintain my speed.
If the light is orange if my speed is average and if the light is off then I brake gently.
If the light is green if my speed is low and if the fire is close then I accelerate

4.3.2 Causal Diagram Representation

The causal diagram of safety behaviour that was developed (see Fig. 4.2) takes several psychological, psychosocial, organizational and economic factors into account. Most current studies of causal relationships between the factors that influence human behaviour are based on the analysis of individual relationships. Here, the factors that have been identified by previous works, and the causal relationships between them were aggregated to build a generic model of safety behaviour.

This causal diagram shows the overall interdependencies between the factors that affect the safety behaviour of operators. The diagram was structured using the factors that directly influence safety behaviour—motivation, stress, self-efficacy, accuracy, and intelligent actions—and those that indirectly influence safety behaviour (by affecting those factors that have a direct effect).

4.3.2.1 Factors That Directly Influence the Safety Behaviour of Operators

Motivation

The operators' level of motivation is directly influences their decision to act. Several theories of motivation have been proposed (e.g., Maslow 1954; Herzberg and Voraz 1959; McClelland 1961; Hackman and Oldham 1976; Vroom 1964; Adams 1963; Apter 1982). Motivation is regarded as a hypothetical construct that is used to

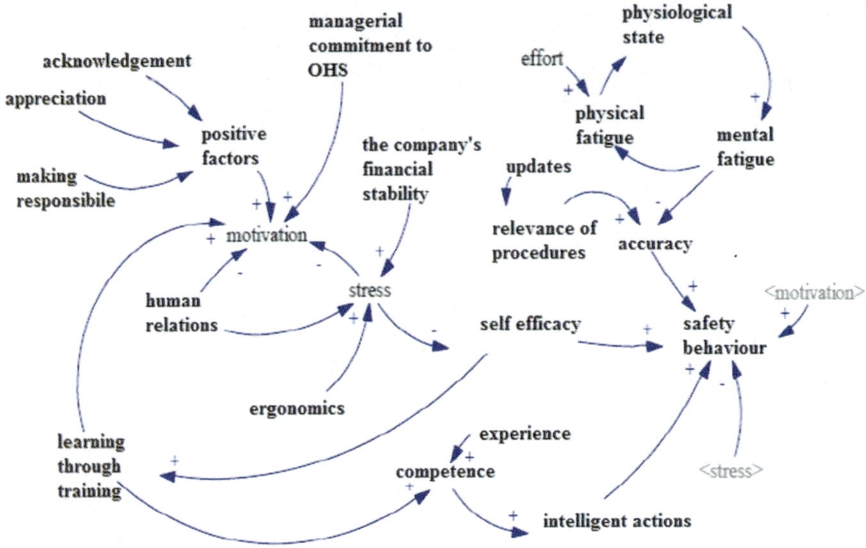

Fig. 4.2 Generic causal diagram of safety behaviour. Note that the direct effects of motivation and stress on safety behaviour are shown at the right hand side of the figure for simplification

describe the internal and/or external forces that lead to the initiation, direction, intensity and persistence of behaviour (Vallerand and Thill 1993). This resonates with the definition used by Karsky and Donnadieu (1990) who developed a detailed model of motivation using system dynamics to explore why a person who is subject to a change of environment decides to act in a particular way.

There are many factors that affect the motivation process negatively and positively. There is a strong link between human relationships and the motivation of operators (Simard and Marchand 1997): aggressiveness and sanctions increase stress and consequently demotivate operators, while recognition and appreciation of their work increases motivation. Simard and Marchand (1997) also showed that organizational factors, such as the degree to which a company's senior managers are involved in occupational health and safety (OHS), positively influences the motivation. This, in turn, influences the operators' safety behaviour.

Similarly, giving operators responsibility—the freedom to take decisions and initiate action—is another way to positively influence motivation. Operator training also provides a source of motivation that increases the commitments of employees to particular activities.

Stress

The workplace remains a major source of psychological stress, even after all other external sources have been eliminated (Marchand et al. 2005). These sources can be down to human relationships—aggression increases the level of operator stress, for example—or the conditions in the working environment, such as noise, vibration, and temperature. Stress must therefore be included in any analysis of human behaviour in the workplace.

Hart and Cooper (2001) represent stress as a process consisting of three major conceptual phases that are influenced by personal, social and environmental factors. The process comprises three major phases: stress factors, stress, and consequences (Harvey et al. 2006). Stress factors are the sources of stress that are found in the organization or the working environment. Stress itself, the second phase of the stress process can be long- or short-term depending on the nature of the stress factors. This second phase constitutes the psychological interpretation and the experience of events that are seen as stressful by an individual. Finally, behavioural, psychological, physiological and organizational consequences appear as a result of prolonged and acute stress.

Stress has been modelled using system dynamics (e.g., Jones 2005; see Fig. 4.3). In Jones' model the causal relations between stress and various psychological factors, such as motivation and mood, are developed.

It has also been suggested that the global financial crisis affected workers' stress levels—a sort of economic stress—Autissier and Bensebaa (2009). Of the 522 respondents in the study, 41% reported an increased level of stress as a result of the financial crisis.

Self–Efficacy

Self-efficacy is an important factor that directly influences operator behaviour. According to social cognitive theory (Bandura 2002), self-efficacy is the belief that an individual has the capacity to carry out particular tasks. People with a greater feeling of self-efficacy set themselves higher goals, and are more committed to their achievement. The trust that a person places in their ability to produce the desired effects influences their aspirations, choices, vulnerability to stress and depression, and level of persistence (Bandura 2002). This self-confidence and self-belief is linked with learning through training, and influences motivation (Foucher and Morin 2007).

Accuracy

The accuracy of work is related to the quality and efficiency of operators' activities. It manifests itself through the number of errors made, and consequently directly influences operator behaviour. The appropriateness of (operating) procedures increases the accuracy of work, while a lack of available information or poor information leads to inaccuracy. Many other factors, such as fatigue, affect accuracy in the workplace.

Fatigue can be considered as a decrease in the activity of a living system, given a constant level of motivation (Scherrer 1967). The physiological phenomenon of

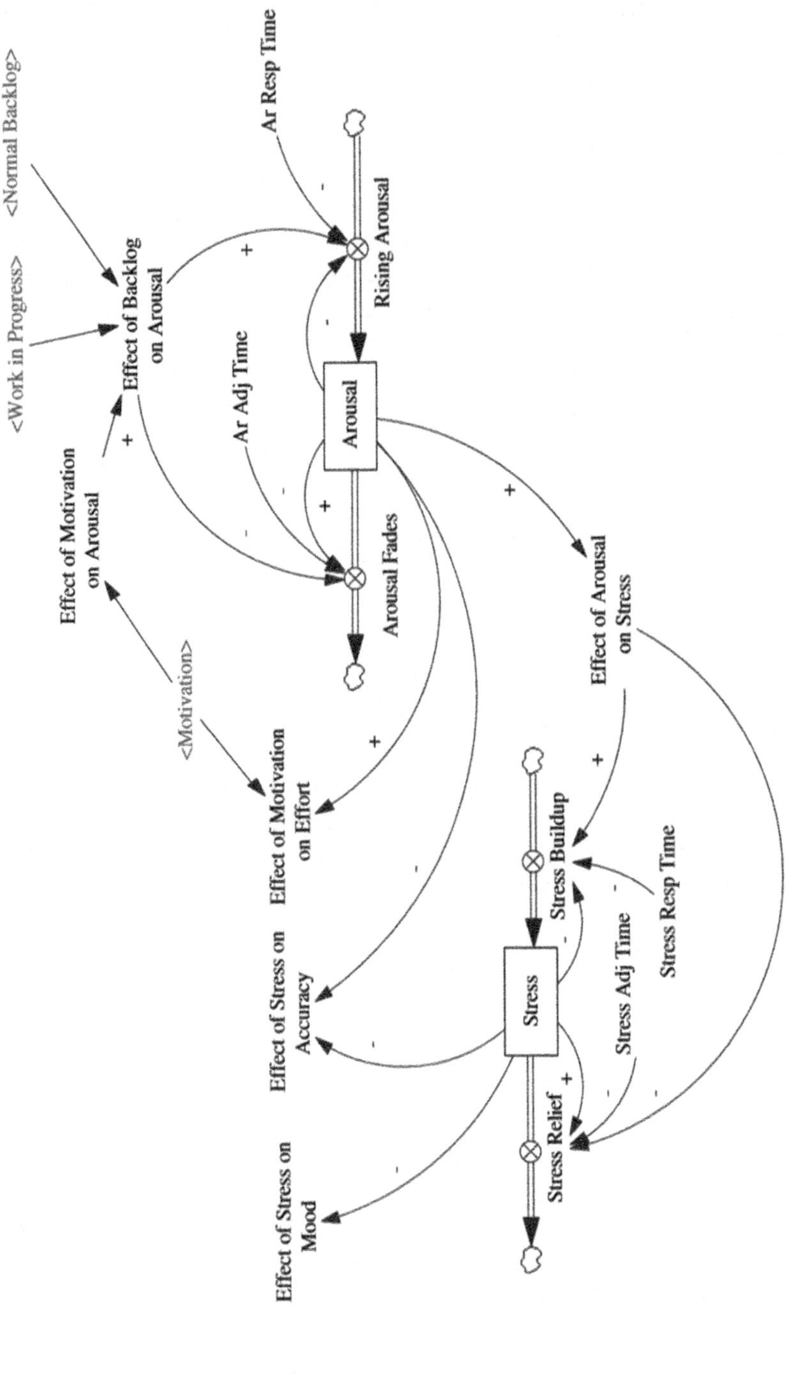

Fig. 4.3 Stress model based on system dynamics. (Jones 2005)

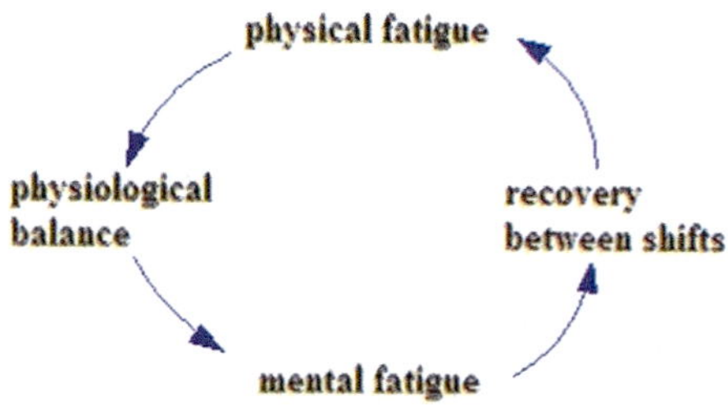

Fig. 4.4 Fatigue loop. (Adapted from Karsky and Adamo 1977)

fatigue can be represented as a simple loop (Karsky and Adamo 1977). This fatigue loop (see Fig. 4.4) is positive and can become dangerous in the event of a nervous breakdown. Mental fatigue manifests itself in an inability to maintain the original plan (Allain 1999): a lack of concentration is accompanied by an increase in errors and reduced vigilance.

Intelligent Actions

Intelligent actions correspond to a mental level that is manifested through the behaviour, adaptation, motor activity, ability to solve problems, and the capacity of the operator to predict the effects of their own behaviour (Gemelli 1955). Intelligent actions are closely associated with individual skills (competence), which is defined as the result of the use of internal expertise to behave in an external context (Zachary et al. 2001):

$$\sqrt{b^2 - 4ac}$$

Several models of competency have been developed (e.g., Le Boterf 2004; Boucher and Burlat 2003; Bonjour and Dulmet 2002; Harzallah and Vernadat 2002).

Competence can be seen as both a process, and as a disposition to act (Le Boterf 2004). As a process, competence can be used to build or adapt strategies for action by drawing upon the resources needed to perform a given task. Through this process, the subject learns and develops their professional practices. Learning through training is one way to develop competence (Cesar 2005) and is a powerful tool that can help establish safety behaviours (Denis-Remis 2007). Experience is another important source of learning and consequently influences competences: Jimenez (1997) suggests that an individual "applies their prior knowledge, created by their previous perceptual experiences and those provided by their culture."

4.3.2.2 The Simulation of the Model

The stock–flow diagram corresponding to the causal diagram (previously shown in Fig. 4.2) is presented in Fig. 4.5. The values of the model's variables were estimated using data from interviews with employees in the chemical industry. The method

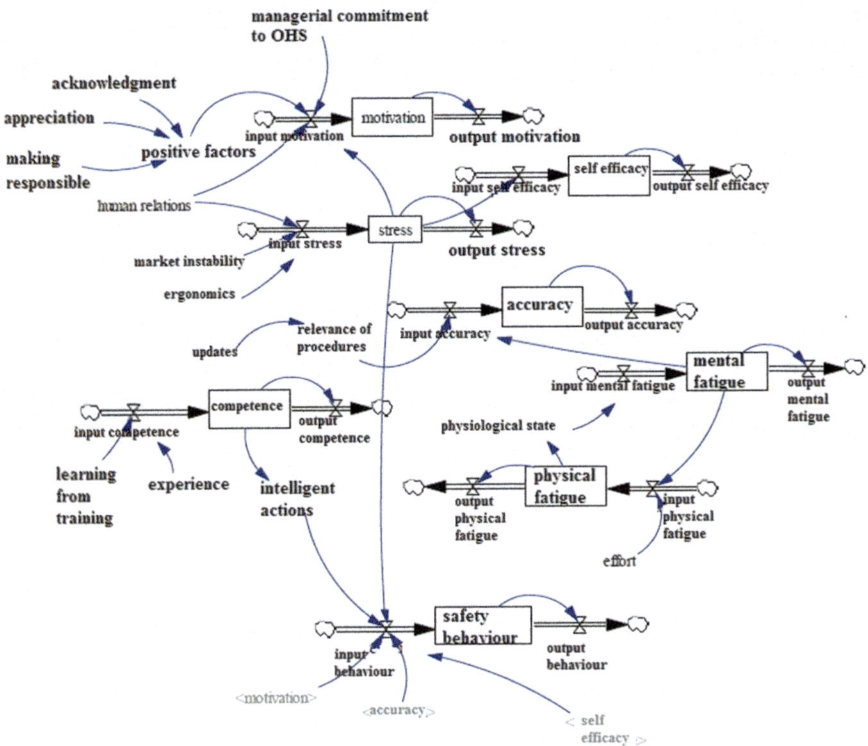

Fig. 4.5 Stock–flow diagram of safety behaviour developed using Vensim® software

involved developing an evaluation grid with a qualitative scale of linguistic values (very low, low, average, high, very high) that were then translated into digital values in the linear range 0–1, where 0 means very low, and 1 means very high.

This grid was presented to the interviewees who were asked to verbally estimate the value of each variable. Sensitivity studies, based on changes in control variables, can be carried out using this model. The following case study explores the impact of motivation on behaviour. The conditions for the simulation are shown in Table 4.1.

The values of the variable *current situation* (of motivation) are estimates taken from the interviews with staff. The simulation period is arbitrarily set to 6 days. The results of the simulation are shown in Fig. 6.13.

Figure 4.6 shows how a change in motivation affects operators' behaviour. As the causal diagram shown earlier (Fig. 6.9) highlighted, several factors (variables) influence motivation. Motivation is a state variable that depends on auxiliary variables such as managerial commitment to OHS, and human relations. A change in the level (value) of one of these auxiliary variables influences motivation and, hence, operator behaviour. This is reflected in Fig. 4.7, which shows the results of a simulation of the effect of a change in the variable *managerial commitment to OHS* on motivation.

Table 4.1 Conditions for the case study

Case Study	Data	Description
Case1: Motivation	*Current situation*	0.75
	High	Increase of 20%
	Low	Decrease of 20%
Simulation period (days)		Start time: 0
		End time: 6
Time step		0.25 days

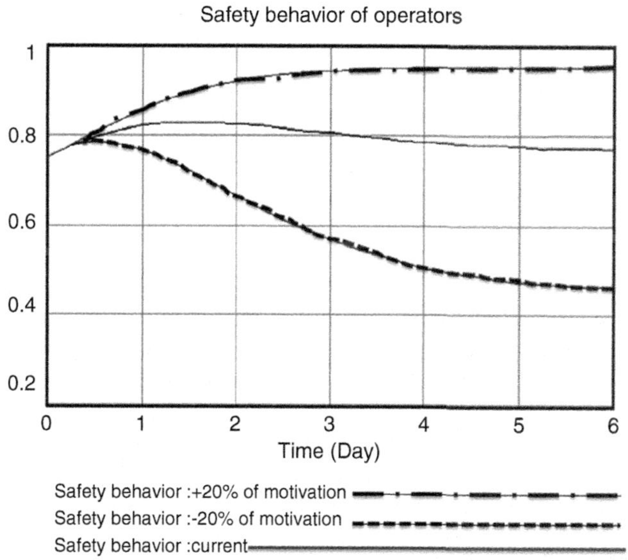

Fig. 4.6 Simulation of the effect of *motivation* on the *safety behaviour* of operators

4.3.3 *Fuzzy Representation of the Model*

4.3.3.1 Use of Fuzzy Set Theory

The development of the causal diagram and description of the relations and causal interactions between the principal behavioural factors highlights the utility of analysing this type of problem using system dynamics. This form of causal model makes it possible to develop linguistic variables whose values are mainly the result of subjective perceptions. Converting linguistic, qualitative, values into numerical equivalents does raise some questions, however. For example, when motivation takes the value of 0.75, the equivalent in natural language is that it is high. What does it mean to say that "Operators are highly motivated"? How can this linguistic value be represented? How can this linguistic quantification be formulated?

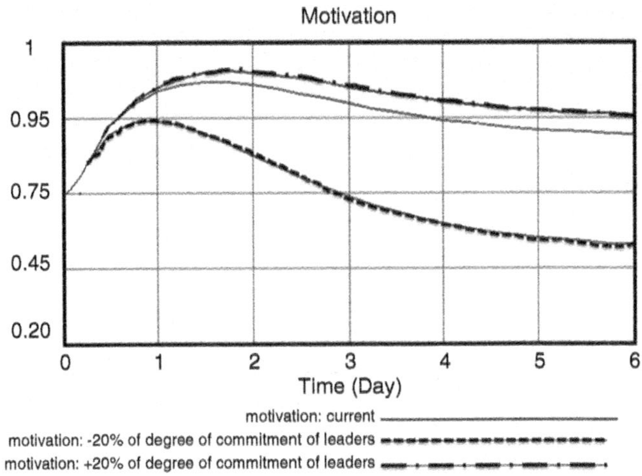

Fig. 4.7 Simulation of the effect of *managerial commitment to OHS* on *motivation*

Another problem lies in the evaluation of certain variables in the model. Responses that begin with "I think that" (e.g., "I think that managerial commitment to prevention is average"; "I think that human relations are poor") suggest there is a level of subjectivity, and hence uncertainty involved in quantifying these variables.

The qualitative nature (e.g., high, low) and the uncertainty (responses that begin with "I think that") associated with these variables needs to be appropriately considered if the model is to be as accurate as possible. To address these issues, the new concept of the *gradual belonging* of an element to a set was introduced into the modelling of the behavioural factors.

4.3.3.2 Application of Behavioural Factors to the Causal Model

The identification of causal relations is clearly defined in the application of system dynamics. The definition and detailed description of these relations made it possible to estimate the strength of each relation in a given loop in the model. The method consists of decomposing the model's feedback loops in terms of individual interaction sequences (as shown in Fig. 4.8).

The estimate of the intensity of each interaction sequence is summarised in a matrix where the axes reflect linguistic values. Each linguistic expression is defined as a fuzzy subset. Figure 4.9 shows an example of a matrix for estimating the intensity of the causal interaction between the variables in the sequence shown above (Fig. 4.8).

This matrix is then interpreted as an interaction. The value of this interaction is derived from the linguistic expression that is suggested by the estimate in the matrix. In the example shown in Fig. 6.15:

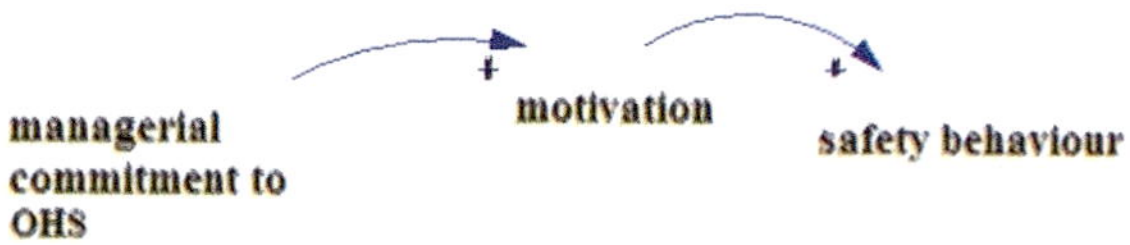

Fig. 4.8 Example of a causal interaction sequence

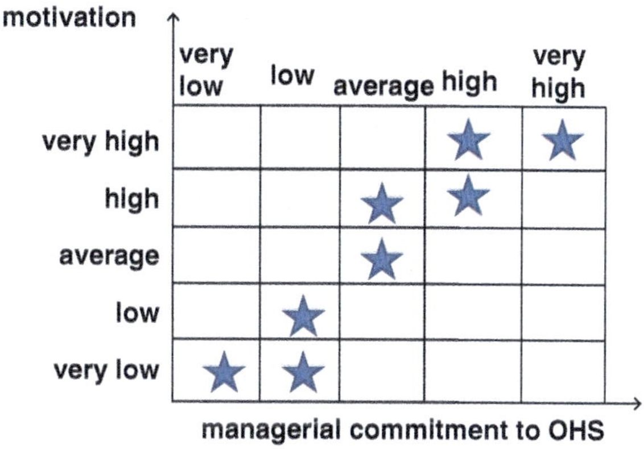

Fig. 4.9 Example of an estimation matrix of the intensity of a causal interaction

- *If* managerial commitment to OHS is very low, *then* the level of motivation is very low;
- *If* managerial commitment to OHS is low, *then* the level of motivation is very low or low;
- *If* managerial commitment to OHS is average, *then* the level of motivation is average or high;
- *If* managerial commitment to OHS is high, *then* the level of motivation is high or very high;
- *If* managerial commitment to OHS is very high, *then* the level of motivation is very high;

The intensity of each of the other interaction in the loops that are shown in the proposed model can be estimated in a similar manner. The evaluation of these relations makes reference to calculation rules found in fuzzy logic.

In this chapter, we have presented a methodology to model and simulate the safety behaviour of operators. This behaviour is considered as a complex system where the uncertainty of the different parameters can be formalized by using the fuzzy logic.

The next chapters present the contributions of an accident model, based on system dynamics called STAMP and introduced by Nancy Leveson in the early 2000, to risk assessment and feedback experience of accidents and project management failures. These examples complete the previous ones, and they illustrate the relevance to apply system dynamics as a key methodology to explore the complexity of socio-technical systems in order to understand and assess risks.

References

Adams JS (1963) Inequity in social exchange. In: Berkowitz L (ed) Advances in experimental social psychology, vol 2. Academic, New York, p 267

Allain H (1999) La fatigue: de la biologie à la pharmacologie. Rapport technique, Université de rennes 1

Apter MJ (1982) The experience of motivation. Academic, London/New York, 378 p

Autissier D, Bensebaa F (2009) Enquête: Impact de la crise financière sur la vie en entreprise. Perception et Comportements des salariés

Bandura A (1997) Self-efficacy: the exercise of control. Freeman, New York, pp 664, ISBN-10: 0716728508

Bandura A (2002) Auto-efficacité. Le sentiment d'efficacité personnelle. Collection "Ouvertures psychologiques", pp 880, ISBN 2744500984

Belyavin A, Farmer E (2006) Modeling the workload and performance of psychomotor tasks. In: Proceedings of the conference on Behavior Representation in Modeling and Simulation (BRIMS), Baltimore, USA

Bezdek JC, Sanker K (1992) Fuzzy models for pattern recognition. PAL, IEEE Press, New York

Bonjour E, Dulmet M (2002) Articulation entre pilotage des systèmes de compétences et gestion des connaissances. 1er colloque de gestion des compétences et des connaissances en génie industriel, Nantes, 12–13 décembre, pp 43–50

Boucher X, Burlat P (2003) Vers l'intégration des compétences dans le pilotage des performances de l'entreprise. JESA 37(3):363–390

Cesar M (2005) The process of organizational learning: to- open access OJSST wards new ways of learning and the Management of Competences

Cornet A (1995) Enquête de climat social et de satisfaction au travail : un outil de gestion des ressources humaines. Rapport technique, 13e Université d'été

Denis-Remis C (2007) Approche de la maitrise des risques par la formation des acteurs. Thèse de doctorat. Ecole des Mines de Paris

Dernoncourt F (2011) Introduction à la logique floue, pp 17

Elkosantini S (2007) Contribution à la modélisation dynamique du comportement d'opérateurs humains. Thèse de doctorat. Université Blaise Pascal, pp 159

Foucher R, Morin L (2007) Sentiment d'efficacité personnelle et apprentissage dans des dispositifs ouverts de formation : résultats de cinq recherches empiriques

Frere R (2000) Contribution à l'intégration d'aspects humains dans la modélisation des systèmes de fabrication : vers une gestion conjointe des ressources humaines et de la production. Thèse de doctorat, Université de Valenciennes et du Hainaut-Cambrésis, Valenciennes, France

Gallo A (1999) Les animaux, psychologie et comportement, pp 63

Gemelli AP (1955) Le Facteur Humain: Des Accidents du Travail dans l'Industrie. 649–660 p

Giambiasi N, Frydman C, Seck M (2005) Modélisation et simulation du comportement humain avec le formalisme devs. In: Actes de 6ème Conférence Francophone de Modélisation et Simulation (MOSIM'06), Maroc

Hackman JR, Oldham GR (1976) Motivation through the design of work: test of a theory. Organ Behav Hum Perform 16(2):250–279

Hart P, Cooper C (2001) Occupational stress: toward a more integrated framework. In: Handbook of industrial work and organizational psychology, vol 2. Sage, Thousand Oaks

Harvey S, Courcy F, Petit A, Hudon J, Teed M, Loiselle O, Morin A (2006) Interventions organisationnelles et santé psychologique au travail: une synthèse des approaches au niveau international. Rapport, IRSST

Harzallah M, Vernadat F (2002) It-based competency modeling and management: from theory to practice in enterprise engineering and operations. Comput Ind 48(2):157–179. https://doi.org/10.1016/S0166-3615(02)00003-9

Herzberg F, Voraz C (1959) Le Travail et la Nature de l'homme, Entreprise moderne d'édition, Paris

Howard PJ (2006) The owner's manual for the brain: everyday applications from mind-brain research, 3rd edn. Bard Press, Austin, p 104
Jimenez M (1997) La psychologie de la perception. Flammarion, pp 125, ISBN 2-08-035496
Jones C (2005) Behavioral theory in simulation: ambiguous results from simple relationships. In: Proceedings the 23rd international conference of the System Dynamics Society, USA
Karsky M, Adamo M (1977) Application de la Dynamique des Systèmes et de la Logique Floue à la Modélisation d'un Problème de Postes en Raffinerie. Actes du Congrès de l'AFCET. Edition Hommes et Techniques. Tome 2, pp 479
Karsky M, Donnadieu G (1990) The dynamic of behavior and motivation. In: Proceeding of the international system conference, Boston
Le Boterf (2004) Construire les compétences individuelles et collectives: La compétence n'est plus ce qu'elle était
Leveson N (2004) Model-based analysis of socio-technical risk, Massachusetts institute of technology, Engineering systems division, working paper series, ESDWP- 2004-08, December 2004
Marchand A, Demers A, Durand P (2005) Does work really cause distress? The contribution of occupational structure and work organization to the experience of psychological distress. Soc Sci Med 60:1–14
Maslow AH (1954) A theory of human motivation. Psychol Rev 50:370–396
McClelland DC (1961) The achieving society. Van Nostrand, Princeton
McCrae R, John O (1992) An introduction to the five-factor model and its applications. J Pers 60:175–215
Novák V, Perfilieva I (1999) Evaluating linguistic expressions and functional fuzzy theories in fuzzy logic. In: Computing with words in information/intelligent systems 1, vol 33. Physica-Verlag HD, Heidelberg, pp 383–406
Scherrer S (1967) Physiologie Du Travail
Siebers P, Baines T, Mason S, Ladbrook J (2004) Modeling human variation in assembly line models. In: Proceedings of the 2004 operational research Society Simulation Workshop (SW04), UK
Simard M, Marchand A (1997) L'adaptation des superviseurs à la gestion participative de la prévention des accidents. Rel Indust 50(3):567–586. https://doi.org/10.7202/051035ar
Vallerand RJ, Thill EE (1993) Introduction au concept de motivation, chapitre Introduction à la psychologie de la motivation, pp 3–39
Vroom V (1964) Work and motivation. Wiley, New York, p 331
Watson JB (1994) Psychology as the behaviorist views it. First published in Psychological Review (1913), 20:158–177
Zachary W, Santarelli T, Ryder J, Stokes J, Scolaro D (2001) Developing a multi-tasking cognitive agent using the cognet/igen integrative architecture. In: Proceeding of 10th conference on computer generated forces and behavior representation, Norfolk
Zadeh LA (1965) Fuzzy sets. Inf Control 8:338–353
Zadeh LA (1975) The concept of a linguistic variable and its application to approximate reasoning. Inf Sci 8(3):199–249
Zadeh LA (1978) Fuzzy sets as a basis for theory of possibility. Int J Fuzzy Sets Syst 1:3–28 978

Chapter 5
Stamp and the Systemic Approach

Karim Hardy and Franck Guarnieri

The desire to overcome the limitations of traditional accident models to take into account all of the interactions that are found in modern socio-technical systems, has led to the development of models that can represent all systemic changes. These models are founded on the idea that accidents are the result of inadequate controls[1] that lead the system toward an accidental state.

The STAMP model of causation (Leveson 2004, 2006, 2011; Hollnagel et al. 2005), which was briefly introduced in Chap. 1, incorporates the concept of inadequate controls. It can be used to identify static and dynamic structures, through tools that can identify, analyse and evaluate any changes in the organization that may lead to an accident. STAMP describes accidents as a systemic change process, rather than a chain of events in which system safety constraints are inappropriately applied in the design, development or operation of a system (Dulac 2007).

This chapter begins with a discussion of the STAMP model's foundations in systems theory (see Chap. 1) and control theory. Next, the three main concepts—constraints, control structures and process models—that form the basis for the STAMP model are described. Finally, two hazard analysis techniques are outlined: the System-Theoretic Process Analysis (STPA) and the Causal Analysis based on STAMP (CAST). These methods, which were developed from the STAMP model, can be used to analyse accidents and to assess the level of system safety.

[1] These inadequate controls translate into application safety constraints that are incompatible with maintaining an acceptable level of safety.

K. Hardy
College of Aeronautics, Worldwide Campus, Embry-Riddle, Aeronautical University, Daytona Beach, FL, USA
e-mail: karim.hardy@erau.edu

F. Guarnieri (✉)
MINES ParisTech/PSL Research University, CRC, Sophia Antipolis Cedex, France
e-mail: franck.guarnieri@mines-paristech.fr

F. Guarnieri, E. Garbolino (eds.), *Safety Dynamics*, Advanced Sciences and Technologies for Security Applications,
https://doi.org/10.1007/978-3-319-96259-7_5

5.1 Theoretical Foundations of the STAMP Model

One of the foundations of the STAMP model is systems theory, which was presented in detail in Chap. 1. Systems theory focuses on the dynamic nature of complex systems:

> *In systems theory, complex systems are modeled as a hierarchy of levels of organization, each more complex than the one below, where a level is characterized by having emergent or irreducible properties. Hierarchy theory deals with the fundamental differences between one level of complexity and another. Its ultimate aim is to explain the relationships between different levels: what generates the levels, what separates them, and what links them. Emergent properties associated with a set of components at one level in a hierarchy are related to constraints upon the degree of freedom of those components. Safety is an emergent property of systems that arises from the interaction of system components.* (Leveson 2004)

The other foundation of the STAMP model is control theory, which is principally based on the concept of feedback. Feedback loops are used in a wide range of settings such as: in nuclear power plants they help to maintain a constant reactor temperature when the ambient temperature varies by several tens of degrees; in aviation, they help aircraft to maintain heading, altitude and speed, and land without any human intervention. They are also deployed by the United States Federal Reserve to help stabilize the economy, through a subtle interplay of feedback, controls and constraints.

Control theory was originally developed as a way of formally explaining the operation of dynamic systems. The intentional use of feedback mechanisms to control the behaviour of a dynamic process is the subject of control systems engineering, and it is this aspect that is generally referred to as control theory.

The remainder of this section explores the foundations of the STAMP model in more detail. The challenges related to control are introduced, then systems control measures are addressed. This is followed by an introduction to the conditions for control, ending with a discussion about control in the context of safety.

5.1.1 The Challenges of Control

The most important justifications for the need to control a system fall into four categories (Bossel 2007):

1. economics;
2. safety;
3. performance;
4. and reliability.

These are discussed in turn below.

Economic considerations are particularly important in the control process, i.e., the control of production systems. Many of these systems, referred to as *continuous processes*, have been designed to maintain a stable state. For example, factories are continuous processes that are designed to permanently maintain stable operating conditions, except when the system is started or stopped. These processes are often regulated to maximise economic yield, and are subject to other constraints that must be applied to variables linked to quality and safety. This is often the case when the optimal solution requires the system to maximize production by operating at its limits.

Safety is a critical dimension of control. Hazardous elements are found in most systems, and the purpose of control is to minimise the effects of those hazards. When taking off, an aircraft has very limited visibility and must be controlled to ensure a safe departure; a nuclear reactor must operate in such a way that the main variables are maintained within safe operating limits.

A given level of performance cannot be achieved without control. An aircraft, for example, must have a system of controls in order to achieve the required level of manoeuvrability. Similarly, a nuclear power plant must perform at a level that can provide energy to an entire population in a way that is safe. These examples highlight that performance is ineffective without adequate controls.

Controls are also used to improve reliability. When faced with unexpected situations, systems are prone to failure. The need for reliability is reflected in the use of redundancy in physical systems to help compensate for any unwanted failures: in the event of one component failing, the second component can (more or less) immediately take over.

Here, *control* means "*to have a deliberate effect on the state of a dynamic system*" (Ogata 1997). In the context of socio-technical systems, the word control is associated with the control of people, found in concepts such as *command and control*. When it is desirable to retain influence over the work to be carried out, controls are applied to the behaviour of individuals or groups, and the concept of command and control is useful, which is not the case when it is necessary to control the safety of a system and the ability of an organization to learn from mistakes.

Control theory seeks to identify the system state variables to be controlled and appropriate ways to control them to maintain system safety, by avoiding any "*a state or set of conditions of a system (or an object) that, together with other conditions in the environment of the system (or object), will lead inevitably to an accident (loss event)*. From this perspective, risk management can be regarded as a control problem, in that the desired (safe) state of the system and the constraints on its behaviour that are needed to keep it in this state can be identified, and a socio-technical control structure can be defined to implement safety requirements (Leveson 2004; Leveson and Dulac 2005). Once the desired, safe system state has been achieved, control is used to maintain that state. The control system translates the system's purpose into actions that are designed to achieve or maintain a goal, through the use of control loops.

5.1.2 *Control Loops*

Control systems can be categorised based on whether they rely upon open loops (using feedforward) or closed loops (using feedback).

5.1.2.1 Open Loops

The open (or feedforward) loop is the simplest control structure. It is usually reserved for applications in which feedback controls are either impossible or ineffective. Its simplicity makes it an ideal starting point for studying control structures and their behaviour, whilst also acknowledging its limitations. Concepts such as the conditions of stability and performance limits take a relatively simple form, but they can become significantly more complex in other control structures.

An open loop system is characterized by outputs that respond to inputs, but are unable to influence them in turn. This means the system is unaware of its own performance (Forrester 1969). In open loop systems, previous actions have no influence on future actions: the controller provides the observer, if there is one, with no feedback about the current state of the system. This means that the controller cannot use real-time information about the system state to tell the system's actuator (a device that acts on the system) which actions to take.

5.1.2.2 Closed Loops

The closed (or feedback) loop is governed by its past behaviour. This system has a closed loop structure that incorporates the results of past behaviour in future decisions. A system is described as a *first order* system if it only contains one state variable; if it contains two, it is a *second order* system. Control systems can be made up of feedback loops (Figs. 5.1 and 5.2) with several fundamental elements (Leveson 2004, 2006, 2011; Leveson and Dulac 2005):

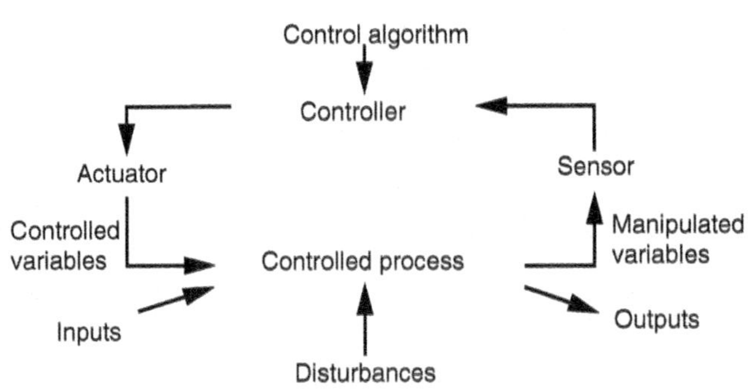

Fig. 5.1 The standard control loop. (Adapted from Leveson 2005)

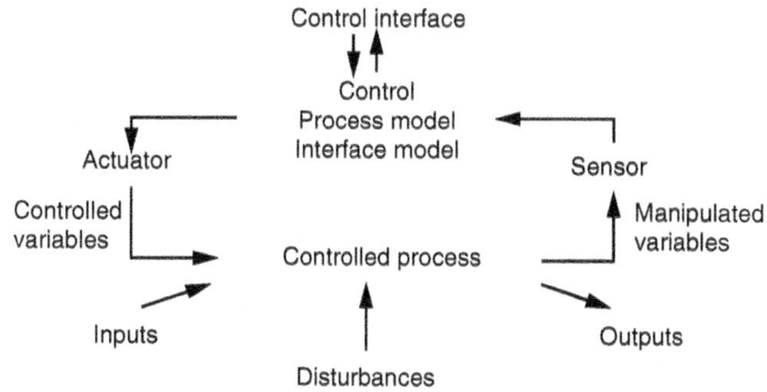

Fig. 5.2 The supervised control loop (adapted from Leveson 2005) wherein the human supervisor may have direct access to system state information (not coming through the computer) and may have ways to manipulate the controlled process state other than through computer commands

- *controllers* that implement the logic of the control system and determine the control actions to be performed. It includes a model of the rest of the system, including other control elements;
- *actuators* that imposes the will of the controller on the system by performing the control action;
- *observers* (or *sensors*) that verify the state of the system.

In the case of a closed loop or feedback control, the observer (sensor) provides information about the system state to the controller. The controller uses this information to update its model and modify how it controls the actuator. The controller compares the current state with the goal state and then uses its model of the system to select a control action that will move the system from its current state towards the desired state.

The communication between system elements, and the control of elements are inseparable concepts in systems theory. The inclusion of constraints on activity at a certain level in a system's hierarchical structure defines the laws of behaviour. In systems, hierarchies are characterized by the control processes found at the interface between levels. In an open system, the controls create a need for communication, as the system is perceived as a combination of interrelated elements that maintain system control and a state of dynamic equilibrium through information feedback loops and controls.

5.1.3 Control Conditions

In general, a control must combine four conditions to produce effects on an entire system (Leveson 2006, 2011):

- *a goal condition*: The controller must have one or more objectives (e.g., maintaining safety constraints in the system);

- *an action condition*: The controller must be able to act on the system state to maintain a process within predetermined limits or safety constraints even when there are internal or external disturbances. If there are several controllers or decision makers, their actions need to be coordinated to meet the goal condition. Uncoordinated actions can lead to accidents when controllers fail to discharge their responsibilities;
- *a model condition*: The controller must incorporate a model of the process. Accidents in complex systems generally occur when there are mismatches between the state in the controller's model of the process and the system's real state;
- *an observability condition*: The controller must be able to verify the state of the system using feedback about the process. The controller requires this information to update its process model.

A controller in an industrial facility obtains information about the status of the process from the measured variables (feedback). This information is used to decide which actions need to be taken to keep the process within predefined limits despite any perturbations that may occur. Generally, maintaining an open system in a stable state requires a set of processes, including an information communication process that regulates and monitors the system.

The effect of a control action is generally asynchronous to the process because of the time delay between the sending of the signal in the control loop and the transmission of the information about that signal. An element may not respond immediately to an external signal, so the process loop may need to include the time needed to receive and respond to variables. Some sensors, for example, are only activated after a certain amount of time has elapsed. These delays slow down the system and increase the effects of the disturbance both within the process *and* on the external response, while simultaneously placing additional requirements on the controller. These delays and interactions often contribute to accidents because they lead to unpredictable variations in performance, which have to be appropriately dealt with to prevent the system from failing.

Accident models based on systems theory consider that accidents arise from interactions between system elements. Unlike industrial safety models (which focus on unsafe conditions), systemic accident models seek to identify the sources of internal malfunctions in the operation or organization of the system, and the potential causes of accidents through interactions between elements.

The systems approach regards safety as a property that emerges where elements interact in an environment. These emergent properties—including safety—are controlled by constraints related to the behaviour of the system elements. Accidents occur when interactions between elements violate these constraints or when there is a lack, or absence of constraints.

5.1.4 Control and Safety

Safety can be treated as a system control problem. Accidents occur when the control system cannot effectively mitigate the effects of a system failure, an external disturbance or a failure in the interaction between system elements. In the Challenger space shuttle accident, for example, the faulty O-ring seal could not properly control the propellant (Leveson 2004); and when the Mars Polar Lander was lost, the software could not control the speed of descent because it failed to discern that the noise in the feedback of the measured variables was indicating that the craft had already reached the planet's surface. These accidents, which were both linked to design errors, were the result of an accumulation of inadequate controls during system development.

When controls are prescribed by an organisation's management, they take a different form. The investigations into the Challenger and Columbia accidents (Leveson 2004) attributed the causes to inadequate controls in the launch process and in system development, along with failures in communication. While events and ineffective behaviours reflect the presence of inappropriate interactions and the inadequate implementation of safety constraints, inadequate controls are the result of events, i.e., of system behaviour. If it is accepted that inappropriate events are due to systemic behaviour in a given context, it follows that these events result from an inadequate control in the system. The control structure has to be examined to determine the inadequacy, i.e., why the constraints on safe behaviour cannot be maintained, and why events occurred. In the case of the Challenger shuttle, the unsafe behaviour translated into an overspill of hot propulsion fuel through the seal on an O-ring: the sealing control function did not perform as expected. In other words, the system was lost because of a failed control during system design.

To understand why accidents happen, the development *and* the operational phases of the system have to be studied to determine the point when a safety constraint was poorly integrated into the system.

In systems theory, which is one of the foundation of systems engineering, systems are regarded as an integrated whole even when they comprise several, specialized elements (which may be other systems, or sub-systems). The main objective of systems engineering is to integrate subsystems into the system as efficiently as possible to achieve the system's purpose as specified in the requirements and design criteria. The optimization of the system design often involves making trade-offs between those design criteria.

A safety approach to systems engineering postulates that certain (emergent) properties of systems, such as safety, can only be understood in their entirety by considering all of the variables, together with social and technical aspects.

In practice, the optimization of a subsystem or elements might not lead to the optimization of the system: improvements to one subsystem may impair system performance because of nonlinear interactions between the system's elements. In addition, the behaviour of individual elements cannot be understood if they are not considered as a part of an overall system, or in terms of their interactions with other system elements.

Traditional attempts to improve safety in the long term have often failed. Any accident model based on systems theory must take account of system engineering principles and the constant need for safety. Models that do not are inevitably limited in their ability to maintain an equilibrium, and consequently prevent accidents. This leads to a lack of synchronization between the system and its environment. Models that *do* incorporate systems engineering principles, however, can improve their own ability to support the development and operation of complex systems in a safe way.

As discussed in Chap 1, modelling socio-technical systems is not new; Forrester (1961) was one of the pioneers of system dynamics; a concept that enabled the development of the idea of the control structure in the STAMP model.

5.2 STAMP Model Concepts

The STAMP model considers safety as a control problem. Control is influenced by many aspects of organisational culture, such as: an authority; company policies, procedures, and standards; regulations; and shared values. It can also be affected by the environment, including the social and organizational contexts in which the system operates. Control is effected through the incorporation of constraints into a control structure with the goal of keeping the system within safe limits, while preserving a clear overview of process models. The STAMP model combines three interrelated concepts: safety constraints; hierarchical control structures; and process models.

5.2.1 *Safety Constraints*

Safety constraints are central to the STAMP model. In systems theory, control always calls for the incorporation of constraints. An accident is not regarded as the result of a series of events, but as a lack, or failure to incorporate constraints at each level of the socio-technical system. Safety constraints are therefore targeted at the relationships between system variables that constitute non-hazardous system states. These constraints are also associated with the control processes that manage the system's ability to change and adapt.

The role of the safety engineer is therefore to identify those constraints that are needed to maintain a safe system state, and to ensure that the design incorporates social, organizational and technical constraints. This has to be done for all levels of the system hierarchy, and for all of the actors in the system.

5.2.2 *Hierarchical Control Structures*

In systems theory, a system is defined as a hierarchical structure in which each level imposes constraints on the activity of the lower level (Leveson 2004, 2006). Socio-technical systems can therefore be modelled as a hierarchy of organizational levels with control processes operating at the interfaces between these levels in order to control processes at lower levels. At each level, an inadequate control may cause the omission, incorrect application or incorrect transmission of a constraint to a lower level. Accidents result from this failure to correctly apply constraints at the different levels of the socio-technical system.

Accident prevention and analysis requires the design of a control structure to represent the context of the socio-technical system. This control structure needs to be as accurate as possible, because it is used during both system development and system operation to ensure that the system complies with its functional requirements. A generic socio-technical structure is shown in Fig. 5.3.

Using the STAMP framework to model how complex socio-technical systems—including organisations—behave in a particular context involves organizing systems into hierarchical levels along with the control processes that act between those levels. The general model in Fig. 5.3 has two hierarchical control structures (shown as A and B). The first (A) relates to system development, while the second (B) covers the operation of the system including both horizontal and vertical interactions (C) between the different hierarchical levels. Operational safety depends partly on the structure of the initial design and its development, and partly on operational control actions. Note that safety is an emergent system property, so it can *only* be improved by adopting a holistic view of the system.

It is important that effective communication channels are established between the hierarchical levels of each control structure (D). The downward transmission of information—from the upper level (D_1) to the lower level (D_2)—allows the incorporation of safety constraints and the application of control; the upward transmission (D_2) provides feedback about the incorporation of the constraints. These exchanges between levels make it possible for each level to assume responsibilities for safety.

For the system development control structures, existing standards, regulations and company policies are tailored to meet system requirements. Any design constraints that are needed to control systemic hazards are passed to the designers and certification bodies as required by the regulations and requirements. At the end of the development process, the results of the risk analysis are incorporated into the design and development documentation that is used to maintain safety levels throughout the system's life cycle (maintenance). Validation of the final system is based on test reports and analyses of hazards and risks.

The process of creating the development control structure is mirrored in the creation of the operational control structure (B). Control structures related to system development also interact with those related to system operation. The safety requirements that are put in place during system development provide a basis for the operational procedures.

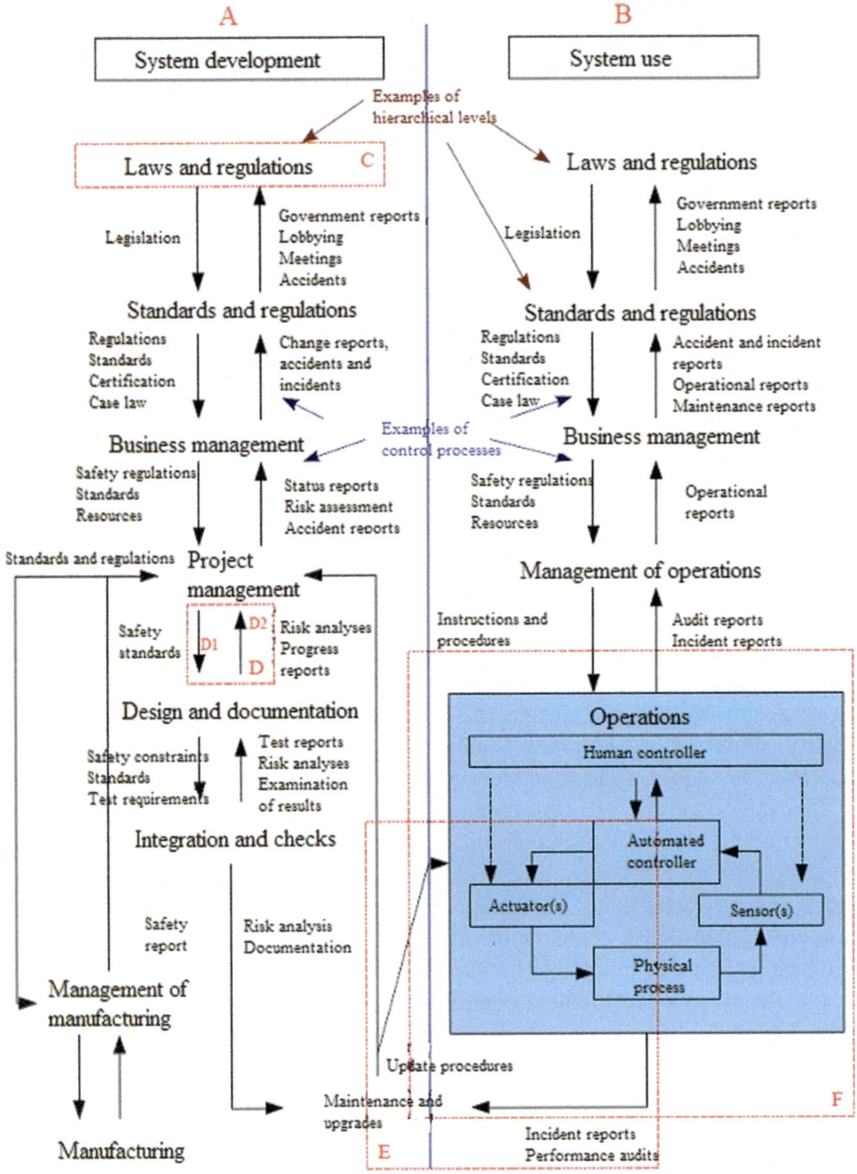

Fig. 5.3 General representation of a socio-technical control structure. (Adapted from Leveson 2005, 2011)

Delays in the transmission of information can affect the flow of control actions and feedback, adversely affecting the control loop's performance and, hence, system safety. Similarly, if the safety constraints on the behaviour of the system elements are poorly implemented, this causes weaknesses in the control loops, leading to accidents. These malfunctions affect interactions between the hierarchical system levels resulting in design flaws, and failures in the elements.

Process accidentology can therefore be regarded as a set of defects that affect the elements of control loops during both the development and operation of the system, i.e., a set of defects that disrupts the interactions between system elements, leading to poor decision-making. The defects can be classified to help analyse and prevent accidents, and to help identify the factors involved in an accident, and their interactions. The STAMP model can be used prospectively to anticipate and analyse possible system hazards and risks, and retrospectively to understand the behaviour of a system that could have led to a particular accident or a loss of equipment or resources.

5.2.3 *Process Models and Control Loops*

The analysis of process models plays a significant role in the classification of defects. A control process (such as F in Fig. 3.3) operates between each level of the system hierarchy, converting an input from one level into a control over another level. The generic process, which can work upwards and downwards, is illustrated in Fig. 3.4. The controller uses input control information, together with the various measures to generate commands. Commands are then sent to an actuator that converts this command into an action through the vector U (Fig. 5.4), which represents the actions of the actuator that affect the controlled process.

The controller's control algorithm is based on an internal model of the controlled process. The controlled process receives inputs from processes at other hierarchical

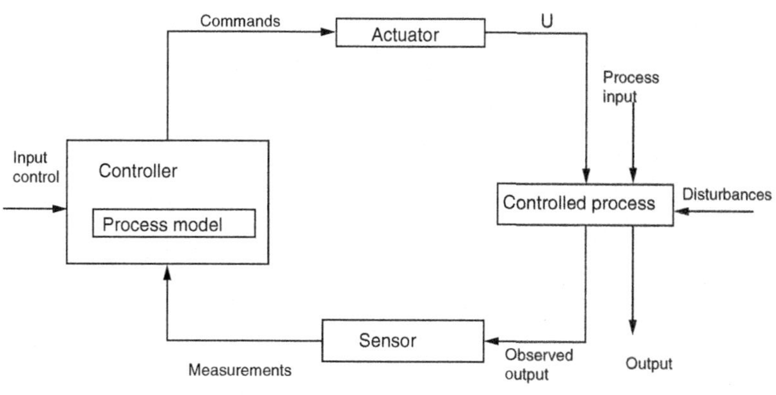

Fig. 5.4 Process control loop. (Adapted from Leveson 2006)

levels and as a result of disturbances. The control input may simply be a goal, a function or a purpose assigned to a system. The outputs of the process can act as a source of information for other connected processes. The controller can send directives to a lower level controller rather than an actuator to control the effect of a process. Sensors measure the process outputs arising from the actions of the actuator and disturbances, and generate new measures that are fed to the controller.

In general, any controller (human or automated) needs a process model to effect control of a system. In all systems—simple and complex—all of the operators and controllers need to include a process model in order to be able to act effectively. This means that, in any process control system, the human controllers (operators) must have a mental model of the process, and the automated controllers should also include a cognitive model of the human operator.

Whether the model is incorporated into a system of automated controls (Fig. 5.5) or human controls (Fig. 5.6), it must contain the same information:

- the relationships between system variables (control laws);
- the current state of the system (the current values of system variables);

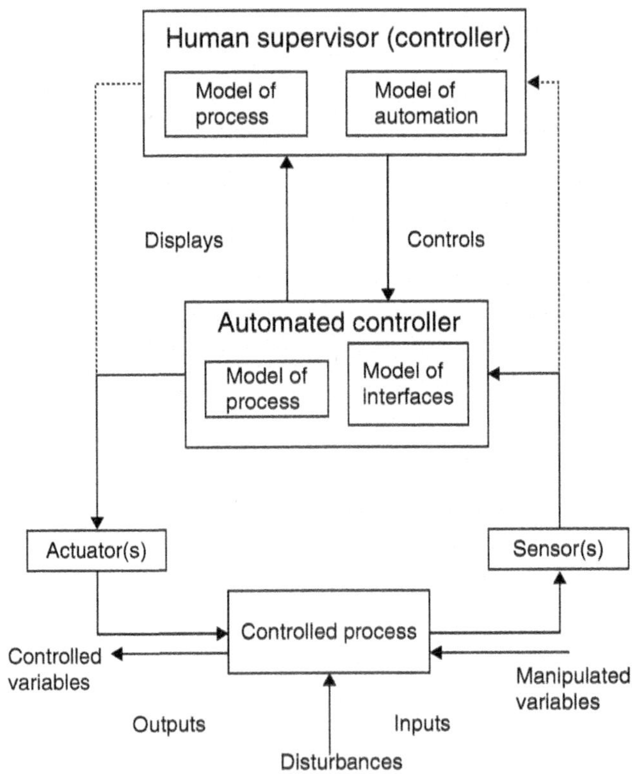

Fig. 5.5 Automated process supervised by a human controller. (Adapted from Leveson 2005, 2011)

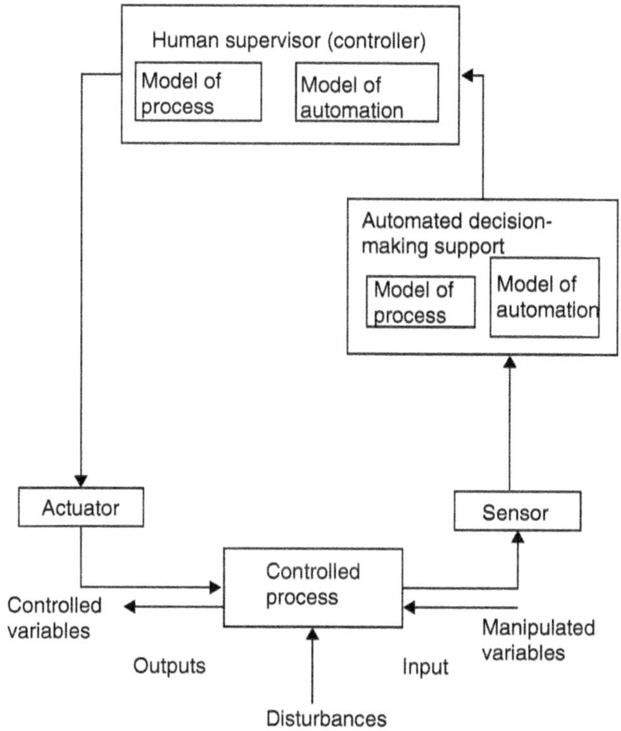

Fig. 5.6 Process controlled by a human operator with automated assistance. (Adapted from Leveson 2005, 2011)

- and the actions that can be taken to change the state of the system.

This model is used to define the necessary controls. The inputs to the model come from feedback loops. This process model is essential at *all* levels in the organisational hierarchy.

In complex systems, control loops connect the hierarchical levels in each control structure. Downward channels provide the information and commands that place constraints on a lower level; upward channels report on the effectiveness of these constraints at the lower level. At each level of the control structure, inadequate controls can arise from oversights, e.g., poor communication of safety constraints, or safety constraints that are incorrectly applied at the lower level. This is why feedback is particularly important when the system is operating. For example, the safety analysis process that generates constraints always includes assumptions about the system's operating environment: when the environment changes, these assumptions are invalidated and the existing controls are rendered inadequate. This discontinuity between the environment and the system may lead to them becoming out of step and hence to inappropriate or even hazardous behaviour.

The control loops that act between the hierarchical levels can take different forms. Fig. 5.5 shows an example of a generic automated controller that is supervised by a human controller. In contrast, Fig. 5.6 shows an example of a generic human controller (operator) who is assisted by automated decision support tools.

The controllers' process models must include the main properties of the sensors, the actuators and the environment, in particular those properties that can cause disturbances in the process. One example is the human–machine interface that provides physical controls, displays and alarms. This provides the means through which the models of automated and human controllers are synchronized. A lack of coordination between the two controllers' models may lead to accidents due to excessive time lags, or incorrect information being provided.

These process models are not only essential to system operation; they are also used during system development. Each controller must have a model of the state of the system to be controlled, a model of the interactions between system variables, and the ways in which processes can change the system state. Accidents—particularly ones with systemic causes—arise when there are mismatches between the system state represented in the controllers' process model and the state of the real system at a given time.

System designers who are also involved in the development and operation of a system can use both the system development model and the system operations model. Being able to use both of these models makes it easier to manage safety constraints within the system, thereby avoiding situations that could lead to accidents. Accidents occur, according to the STAMP model, when there is a failure to respect one or more safety constraints, causing a failure in the system elements due to an environmental disturbance or a failure in the interactions between elements (Leveson 2011).

In each control loop, and at each level of the socio-technical system's control structure, an unsafe behaviour translates into either an oversight, or an inadequate constraint on a lower-level process, or the inadequate application of a constraint (Table 5.1). Each element in a control loop is involved in applying safety constraints. These constraints can be categorised begins by assessing the potential contribution of each element in the main control loop: the controller can initiate actions that are correctly performed by the actuator; alternatively, an inadequate or inappropriate control action may be performed, leading to failures or disruptions in the physical process These general factors can apply within *and* between the levels in the socio-technical control structure.

These three concepts (constraints, control structures and the process model) provide the theoretical foundations for the STAMP accident model and accident analysis tools. These are described in the next section.

Table 5.1 Classification of control errors leading to hazards

1. Non-compliant application of constraints (control actions)
(a) Unidentified hazards
(b) Inappropriate, ineffective or overlooked control actions for an identified hazard
(i) Development of the control algorithm without the application of constraints
1. Overlooked in the development of the process
2. Modifications to the process, but no modification to the control algorithm
3. Modification or incorrect adaptation
(ii) Incoherent, incomplete or incorrect process models
1. Weakness in the creation of the process
2. Weakness in the update to the process
3. Inadequate or overlooked feedback
(a) Not included in the system design
(b) Poor communication
(c) Deadlines
(d) Poor sensor operation
(4) Deadlines or imprecise measurements
(iii) Poor coordination between controllers and decision-makers
2. Non-compliant execution of a control action
(a) Poor communication
(b) Inappropriate action by the actor
(c) Deadlines

Adapted from Leveson (2005)

5.3 CAST and STPA Hazard Analyses

This section describes two analysis methods developed from the STAMP accident model: the Causal Analysis based on STAMP (CAST; Leveson 2011); and the System-Theoretic Process Analysis (STPA; Leveson et al. 2003, 2004; Dulac and Leveson 2004; Stringfellow et al. 2007). CAST is used to investigate accidents using static and dynamic perspectives, while STPA is used for safety assessment.

5.3.1 The Accident Analysis[2]: CAST

In STAMP, the system is viewed as a set of interacting control loops, and accidents arise through inadequate control. Using STAMP to analyse accidents is an iterative process that involves examining the socio-technical system to identify weaknesses in its safety control structure and highlight changes to the system.

Every analysis begins with the identification of system hazards, which are translated into strategic safety constraints. The next step defines the safety control structure highlighting the controls and the feedback in the system. This safety control structure is then used to guide the investigation of each hierarchical control to assess its impact. The identification of inadequate control actions highlights the safety constraints that have been inadequately applied. Once the hazardous control actions have been identified, a recommended course of action to deal with them is drawn up.

Both static *and* dynamic models are generally needed when investigating an accident in a particular system. These two models are used to understand the system and validate any changes that must be made to the safety culture of an organization. The models can be used to evaluate and analyse the causes of accidents and detect the accident arose because safety levels had become unacceptable. They can also be used to assess the potential impact of any changes and decisions that may have affected the system's structure, causing it to migrate to an accident state.

The *static* model of safety control is used to visualize the organization of the system, together with the interactions within the system:

- safety requirements and constraints;
- failed control actions;
- the context (social, political, economic, environmental, etc.) at the time of the accident;
- defects in the cognitive models of the actors in the system;
- coordination, communication and interaction defects between actors in the system.
- The *dynamic* model describes the behaviour of the system at the time of the accident, together with how it migrated from a safe state to an accident state. The static and dynamic models are used during the two phases of an accident investigation, in STAMP: the static phase and the dynamic phase. These phases are shown in Fig. 5.7.

[2]Various accident analyses based on the STAMP accident model have been published (Leveson 2011; Leveson et al. 2003.

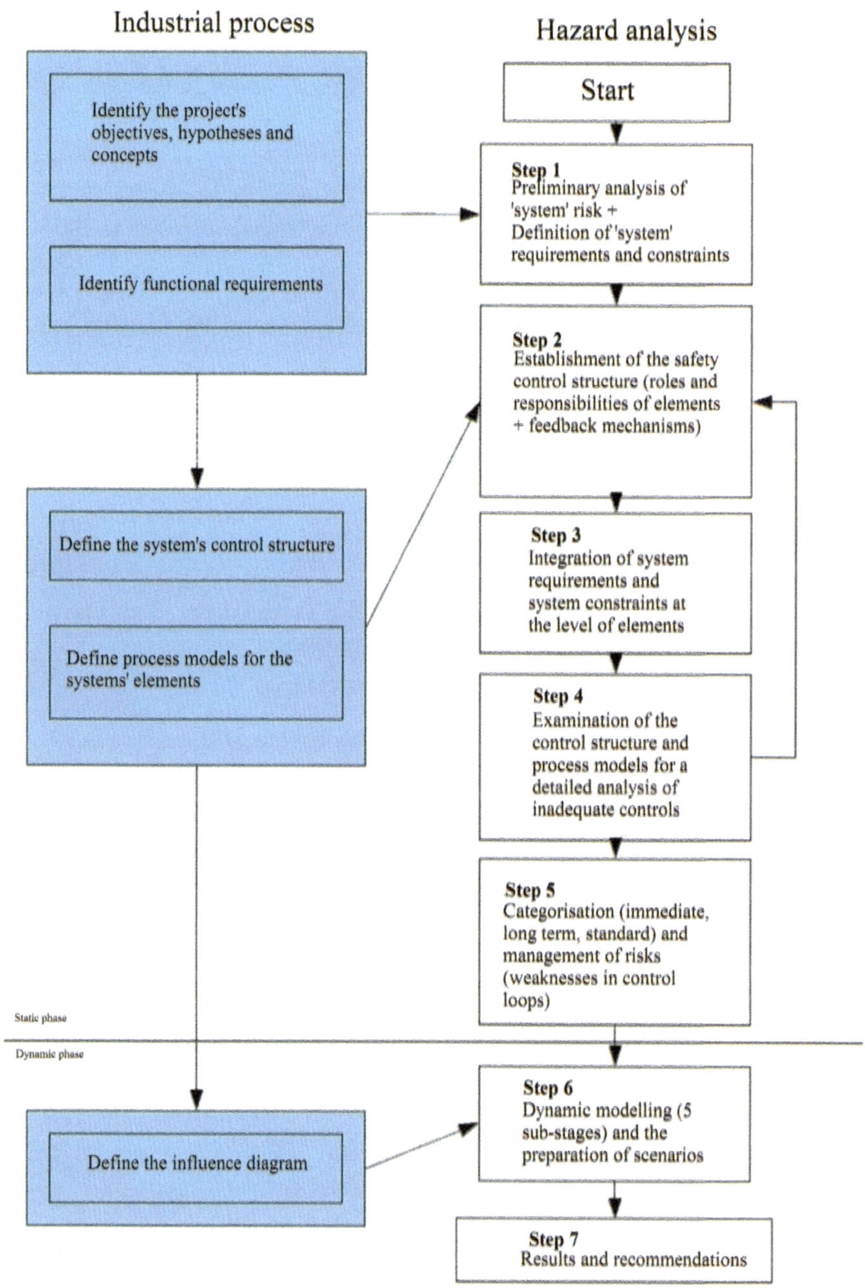

Fig. 5.7 The STAMP process.

5.3.1.1 Static Phase

The static phase of the analysis involves five steps, as shown in the right half of Fig. 5.7:

1. A preliminary analysis of system risk and the definition of system requirements and constraints;
2. The establishment of the safety control structure (roles and responsibilities of elements and feedback mechanisms);
3. Integration of system requirements constraints at the level of the element;
4. A detailed analysis of the control structure and process models to identify inadequate controls;
5. The categorization (immediate, long term, ordinal) and management or risks (weaknesses in control loops).

The steps are described in detail below.

Step 1: Preliminary Risk Analysis and Definition of System Requirements and Constraints

OBJECTIVE—Identify the system's safety requirements and constraints by carrying out a preliminary review of the risks and hazards at the system level. This helps to highlight hazards that have not previously been identified.

METHOD—As with any hazard analysis these requirements and constraints come from the examination of possible failures, including at the level of interactions, and conditions that make it impossible to maintain control of the system.

System hazards relate to the whole system. They can be defined as "engineering failures and a poor decision-making system leading to the accident" (Leveson 2011). System hazards can be used to define the safety requirements and constraints at the system level:

- safety must be a priority in the design and decision processes;
- decisions on safety issues must be addressed by experts;
- safety analyses must be carried out as early as possible in the system's life cycle and the results made available.

Chapman and Ward (2003) developed a taxonomy of the different types of hazards that may be present at the system level:

- hazards related to the system requirements based on estimates of time, cost and quality, rather than on real figures. These hazards can take the form of:
  - a lack of clarity in the definition of design/development requirements. Here the risk is improper system operation leading to the accidental state;
  - a lack of control in a domain, leading to an inadequate control action;
  - unexpected events that may cause operational losses (see, for example, Perrow's (1999) concept of complex interactions)

- hazards related to the design and logistics of the system (was the design process respected?);
- hazards related to the system's objectives. These hazards may imply design modifications or significant compromises. Here it is important to clearly define responsibilities for each system element (see Step 2);
- hazards that appear before the occurrence of an accident due to interactions between the different system elements. These interactions can be complex and produce inadequate controls due to:
  - an incorrect definition of the responsibilities of elements (see Step 2);
  - an inadequate understanding of the role and responsibilities of elements;
  - inadequate communication between elements (see Step 4);
  - the functionality of elements;
  - an inadequate understanding of documentation related to development and operations;
  - inadequate coordination and control mechanisms in the system.

Step 2: Establishment of the Safety Control Structure.

Objective—To construct the system's safety control structure (see Fig. 5.4) including the roles and responsibilities of each element, the control elements, and the feedback loops for each of these elements. This step makes it possible to determine and establish the system's safety control structure (Leveson 2004, 2011). The roles and responsibilities are used to help ensure that the system's safety constraints are appropriately applied.

Method—The first step is to define which elements should be included as part of the system, and which should be excluded. The process is somewhat similar to that of defining the boundaries of a problem during dynamic modelling. It is difficult to include those actors that may have an indirect impact on the system in this analysis. Establishing the scope of the analysis involves making some trade-offs: if a broad scope is used, it becomes more likely that certain factors are identified that would otherwise be overlooked without further analysis. As systems evolve over time, the elements that are included in the analysis should also evolve to reflect the structural changes that can significantly affect the system's dynamics. Structural impacts can affect behaviour, which itself causes structural changes and may lead to different interpretations. Both the model and the analyses must be part of a cycle that resembles a safe process in systems in which hazards and analyses should be revised, updated and clarified.

Dulac (2007) defines a non-exhaustive list of control structure elements that can be used to facilitate the development of a safety control structure. This list also supports, in a second step, the construction of a dynamic model of the system (Table 5.2).

It is useful to specify the inputs and outputs, roles and responsibilities, potentially inadequate control actions and requirements in terms of feedback. It is also useful to include environmental factors and the various controls that are in place. During this step it is essential to identify oversights, and highlight the different responsibilities and any conflicts between them. The inclusion of responsibilities

Table 5.2 List of elements in a control structure

System development	System operation
Laws and regulations	Laws and regulations
National and governmental agencies	National and governmental agencies
Industrial associations	Industrial associations
Consumer associations	Consumer associations
Insurance companies	Insurance companies
Business management	Business management
Project management	Operations' management
Development engineering	Operational engineering
Safety development analyst	Occupational risk specialist
System upgrade	System maintenance
Checks (safety, quality, etc.)	Operational checks (quality, safety, etc.)
Manufacturing	Supplier(s)
Contractor(s)	Contractor(s)

Aadapted from Dulac (2007)

complements the analysis by highlighting weaknesses in the definition of the functional requirements placed on certain elements. This step can be based either on the existing organizational design (if it exists), or on a new design that meets the requirements and constraints of the system. By taking into account the roles and responsibilities of each element it becomes easier to understand the control structure.

Step 3: Integration of System Requirements and Constraints

Objective—To integrate the system requirements and constraints defined in Step 1 at the level of each element in the safety control structure defined in Step 2.

Method—This step defines safety constraints for all of the elements of the safety control structure that are consistent with the constraints defined in Step 1, while simultaneously including potential interactions. These requirements and constraints correspond to the safety responsibilities for each system element.

Requirements and Safety Constraints

The most recent safety rules and regulations must be applied at the system level in order to preserve the safety of employees at managerial level in a business.

Step 4: Review of the Control Structure and Process Models

Objective—To identify any inadequate controls that could contribute to an accident by analysing the control structure.

Method—Using the control structure from Step 2, and the responsibilities and feedback mechanisms identified in Step 3, any inadequate controls can be identified. The inadequate controls fall into four categories:

- a control action that has not been performed;
- an incorrect or unsafe control action that was performed leading to a loss;
- a potentially correct control action that was performed at the wrong time (too early/too late);
- a correct control action that was stopped too soon.

The inadequate control actions are formulated for each hierarchical level. These control actions are a fundamental resource for understanding the system's behaviour.

Example Applied to Fig. 5.3
The hierarchical level "business management" does not provide safety standards to the lower level. This inadequate control may be characteristic of a lack of communication between hierarchical levels.

Step 5: Risk Categorization and Management

OBJECTIVE—To categorise the identified risks in order to determine the impact of inadequate control measures on system behaviour, and to address any control deficiencies by identifying the processes that led to the violation of one or more safety constraints.

METHOD—The risks can be divided into three categories:

- risks to be managed immediately: these risks characterize the hazards that have an immediate impact on the system's safety;
- risks to be managed in the long term: these risks characterize hazards that do not have an immediate impact on the system's safety, so they can be addressed in the longer term;
- ordinal risks: these risks do not require any special or urgent management.

Events resulting from inadequate control actions can be interpreted in terms of defects in the development and operation of the system. These anomalies are found in the control loops (Fig. 5.8) between the hierarchical levels, and lead to constraints being omitted or inadequately applied, which results in the violation of safe behaviour. The aim is to incorporate inadequate control actions that could cause a migration to an accident state into control processes. In addition to the constraints defined in Step 4, the analysis of the control loop and process models highlights the impact of these actions on the behaviour at the hierarchical levels and the consequences that may reverberate to other levels, upwards or downwards.

Overall, the static phase (Steps 1–5 above) leads to the identification of inadequate control actions and of control defects in the process. Whether this is done at the level of interactions between hierarchical levels (Step 4) or within control processes (Step 5), it helps to define potential scenarios that can be used to explain the evolution of the system (in the dynamic phase).

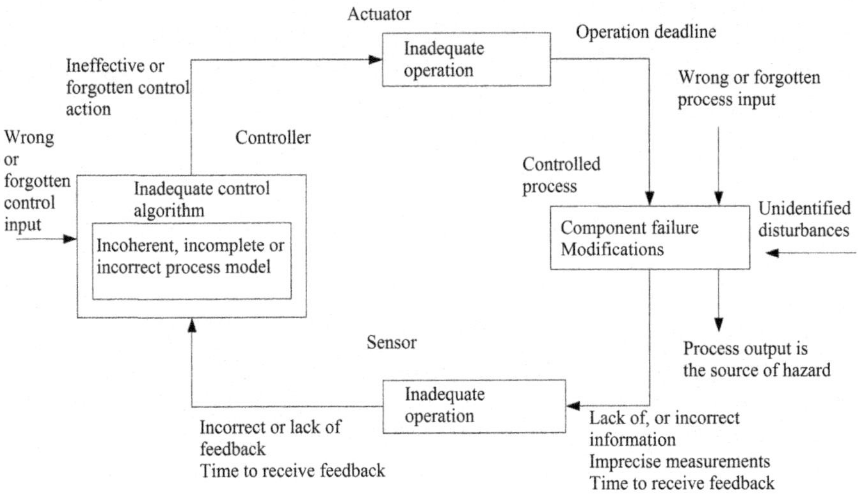

Fig. 5.8 Inadequate control loop. (Adapted from Leveson 2009)

5.3.1.2 Dynamic Phase

The dynamic phase involves two steps:

1. developing a dynamic model of the system (Step 6);
2. formulating the results and recommendations (Step 7).

Step 6: Dynamic Modelling
Dynamic modelling is used to identify the relationships between elements and to better understand system behaviour.

Objective—To identify risk indicators and establish new types of analyses using a dynamic model of system behaviour. This step is essential in the investigation and analysis of an accident. Software tools[3] can be used to simulate behaviours and generate scenarios to help determine the causes and the variables that may have played a key role in the occurrence of an accident.

Method—The construction of the dynamic model uses a process adapted from Sterman's (2000) method, which is used in the field of system dynamics (Table 3.3). The model represents the system at the time of the accident. It can therefore be used to identify the dynamic behaviours that led to the migration of the system towards the unsafe state. The model is validated in an on-going, iterative manner.

The dynamic model is developed in detail and each new piece of information that is derived from interacting elements is added. Three types of information are normally used to create a dynamic model (Forrester 1992): numerical data; textual

[3] For example, Vensim® (http://www.vensim.com) and Stella® (http://www.iseesystems.com/)

Table 5.3 Steps in the dynamic modelling process

1. **Definition of the system**
What is the accident? Why is it an accident?
What are the variables involved in the accident?
On what timescale is the accident investigated?
What was the behaviour of the system at the time of the accident?
2. **Establish the dynamic hypothesis of the accident (its evolution)**
What are the current theories on the evolution of the accident?
Develop a dynamic hypothesis explaining the evolution of the system towards the accident through the interplay of feedback.
3. **Establish a dynamic model**
Characteristics of the accident structure and decision rules.
Estimate parameters, behavioural relationships and initial conditions.
Test to verify the coherence with the objective and the limits of the accident.
4. **Test**
Compare with the baseline.
Robustness to extreme conditions
Any other useful tests.
5. **Develop a strategy for change**
What environmental conditions may appear?
What new rules, strategies or structures can be integrated into the real world?
What are the effects of the new policy?

Adapted from Sterman (2000)

sources; and designers' cognitive models (which Forrester regards as the most important source of information).

The creation of a dynamic model thus requires five steps (Table 5.3). The first is to define the system based on the control structure that was established in Step 2 of the static phase. In this step the systems' base parameters, such as its lifetime, variables, etc., are defined. The coupling between system elements (Perrow 1999) and the impact of changes on other parts of the system are also defined, along with complex interactions, which can lead to unexpected events (Perrow 1999), which influence the occurrence or even the consequences of major accidents in complex systems (see Step 1). If an event is unanticipated, the control actions in the feedback loop will be inadequate, because they cannot deal with it.

It is also necessary to take into account the severity of accidents and the time needed to return the system to normal operation following an accident (or loss) that affects the system's dynamics. In this context, taking into account information from the static phase is also an important part of the characterization of the system, together with uncertainty in political or strategic decisions.

The second step defines the dynamic accident hypotheses. These hypotheses describe the changes that can occur in the system through the interplay of feedback, based on the safety control structure.

In the third step, the structure of the dynamic safety model is developed. This dynamic model is supplemented by information about the many pressures, disruptions and external influences on the system, as well as information about the safety control structure.

These pressures and influences include issues such as:

- production, turnaround time;
- the degree of integration of safety standards and regulations;
- strengthening safety;
- pressure on human and other resources;
- the availability and use of incident and accident reports;
- performance reports;
- cost reports.

The fourth step validates the behaviour of the system elements given its operating limits. This test of the limits is usually achieved through a loop that tests system safety as a function of its context. These tests are designed to ensure that the response of the socio-technical system elements is consistent with the behaviour of decision makers in any given context; they also help to highlight elements that could trigger a migration towards an accident state.

These steps are designed to improve the decision makers' cognitive model so that they understand exactly which safety thresholds must not be exceeded. More generally, the main objective here is the design and operation of a system from the point of view of an accident, to highlight the erosion of safety that led the system to migrate towards the accident state. It is not always possible to anticipate all of the potential increases in risk at the outset. However, many of the organisational risks identified during analysis can be eliminated from the system. This fifth step also requires feedback from experience.

Step 7: Results and Recommendations

Objective—To present the results of the analysis and make recommendations that foster a policy of change within the system. This can be translated into a process that employs experience-based feedback to incorporate the changes into the operational documentation to redefine the roles and responsibilities of all elements. This should lead to new organizational procedures that change the structure and therefore the behaviour of the system.

Method— The dynamic phase helps to restore the behaviour of the system by defining a dynamic hypothesis that can be used in a dynamic model. This phase makes it possible to trace the evolution of the system and to highlight the elements that played critical roles in the accident.

5.3.2 The Safety Assessment: STPA

Hazard analysis is one approach that can be used to help prevent accidents. Current hazard analysis techniques, such as those found in operational safety, however, do not take appropriate account of the dynamic aspects of modern, complex systems where human-machine interactions are important.

An alternative, proactive approach to accident prevention, based on the STAMP model, can also provide the necessary information for risk prevention and therefore accidents. The System-Theoretic Process Analysis (STPA) offers another way to collect information intended to sustain safety-oriented design and development. The objectives of the STPA hazard analysis are similar to those of a traditional hazard analysis:

- to identify hazards throughout the system life cycle, together with the associated safety constraints needed to maintain an acceptable level of safety;
- to guide how these safety constraints can be violated, and how they can lead to inappropriate actions that push the system towards an accidental state.

There are five steps in the STPA hazard analysis process (see Fig. 5.9), which are methodologically similar to the CAST accident analysis:

1. Preliminary analysis of system risk and the definition of system requirements and constraints;
2. Establishment of the safety control structure (roles and responsibilities of elements and feedback mechanisms);
3. Integration of system requirements and constraints at the element level;
4. Review of the control structure and process models and a detailed analysis of inadequate controls;
5. Categorization of risks (immediate, long term, ordinal) and management (defects in control loops).

In accident investigations, the STPA hazard analysis process places great emphasis on the dynamic model and dynamic simulations of the system. In doing so, it aims to maintain coherence between the system state and the actual state. This results in a system that remains in step with its environment to avoid any migration to an accidental state. Unlike traditional analytical techniques that are typically static and focus on the system's ability to avoid unstable states from the point of view of the current system design and its environment, the STPA technique starts from the point that systems are dynamic by nature, and that they adapt and evolve throughout their life cycle.

The STPA analysis looks at the socio-technical system in which human and organizational factors are fully taken into account, as well as technical factors. The presence of the human factor implies both behavioural uncertainty and a lack of information. Uncertainty leads to additional work being needed in the search for information to incorporate into the dynamic model.

The analysis begins during the early phases of system development and continues throughout the life cycle (Ishimatsu et al. 2010). Its use during the design phase allows the incorporation of a safe design process in which the results of hazard analyses can influence design decisions, which become clearer and more complete as new information becomes available.

The STPA analysis, which is based on the principle of a control process loop, takes into account software and hardware aspects, decision processes, and the organizational factors involved in accidents. It reflects its roots in control theory,

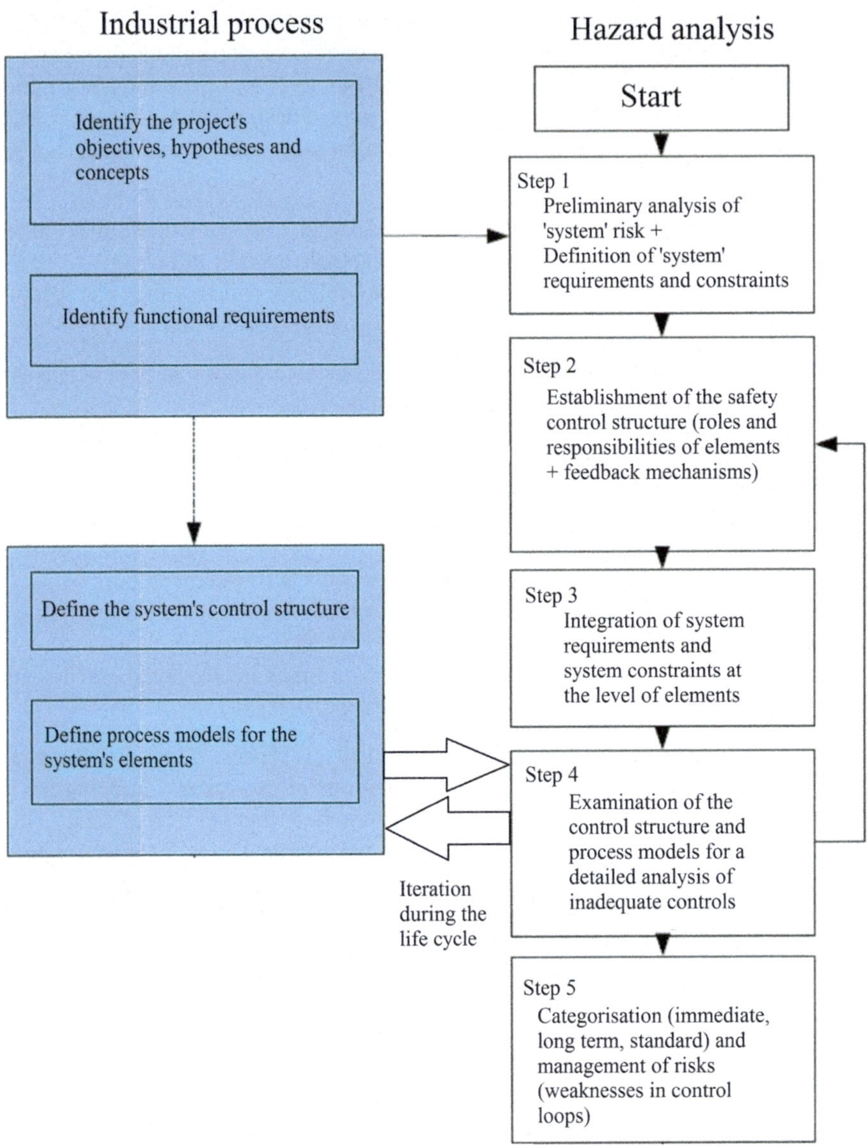

Fig. 5.9 The STAMP process applied to safety assessment. (Adapted from Leveson et al. 2004)

focusing more on controlling hazards than eliminating defective elements, which are just one possible source of hazards.

The analysis identifies the safety constraints that must be applied, together with the controls that implement those constraints and any reasons why the control might be inadequate. This information supports decision making at every stage of the design process, from the earliest phases of the system's lifecycle. The STPA analysis evolves as the design progresses. Even though control requirements cannot be fully defined during system design, they act as a reference point for the design and implementation of safety requirements in the system's operational phase.

A comprehensive hazard analysis has to identify the potential changes to safety controls over time that could lead to insecure states or even accidents. These can be illustrated using influence diagrams. The information from the analysis can be used either to forecast changes during the system design phase; or to identify measures to detect system degradation during its operation.

The STPA hazard analysis technique is applied both iteratively and opportunistically. Engineers can either apply a particular technique, or apply STPA to a wider range of hazards. In the earliest design stages, some decisions can be taken, but control defects or the execution of inadequate controls cannot be identified. However, the STPA hazard analysis, if applied early in the life cycle of a system, provides information to the design process on the system state and its level of risk.

The STPA taxonomy (see Table 5.1) can be used to identify inadequate controls that could lead to inadequate control actions. This information can be used to create new safety constraints, complement existing constraints, or create a new design so that hazards are managed, controlled or eliminated. A design is considered safe or safe enough when the level of safety control is considered acceptable.

The next chapter introduces an example of the application of STPA in the case of an innovative polluted sediment decontamination process.

References

Bossel H (2007) Systems and models, complexity, dynamics, evolution, sustainability. Books on demand GmbH, Norderstedt

Chapman RJ (1998) The role of system dynamics in understanding the impact of changes to key project personnel on design production within construction projects. Elsevier Pergamon Int J Proj Manag 16(4):235–247

Chapman C, Ward S (2003) Project risk management, processes, techniques and insight. Wiley, Hoboken

Dulac N (2007) A framework for dynamic safety and risk management modeling in complex engineering systems. MIT, Cambridge, MA

Dulac N, Leveson NG (2004) An approach to design for safety in complex systems. INCOSE04, Toulouse

Forrester JW (1961) Industrial dynamics. MIT Press, Cambridge, p 464

Forrester JW (1969) Urban dynamics. Productivity Press, Cambridge, MA, p 329

Forrester JW (1992) From the ranch to dystem dynamics: an autobiography. In: Bedeian AG (ed) Management laureates: A collection of autobiographical essays, vol 1 of 3. JAI Press, Greenwich

Hollnagel E, Woods D, Leveson NG (2005) Resilience engineering: chronicling the emergence of confused consensus. Ashgate Publishing, Aldershot

Ishimatsu T, Leveson N, Thomas J, Katahira M, Miyamoto Y, Nakao, H (2010) Modeling and hazard analysis using STPA, MIT, Japan Aerospace Exploration Agency

Leveson N, Dulac N (2005) Safety and risk driven design in complex systems of systems. 1st NASA/AIAA Space Exploration Conference, Orlando, February 2005
Leveson N (2004) A new accident model for engineering safer systems. Saf Sci 42(4):237–270
Leveson N (2009) Engineering a safer world, system safety for the 21st century, Massachusetts Institute of Technology, July 2009 (a draft book). Accessed 1 Dec 2011. http://sunnyday.mit.edu/safer-world.pdf
Leveson N, Daouk M, Dulac N, Marais K (2003) Applying STAMP in accident analysis. Working paper series. Massachusetts Institute of Technology, Engineering Systems Division, Cambridge, MA
Leveson NG (2005) Safety in integrated systems health engineering and management. Paper presented at the NASA ames integrated system health engineering and management forum
Leveson NG (2006) A new approach to system safety engineering. Cambridge, MA, Unpublished Manuscript
Leveson NG, Daouk M, Dulac N, Marais K (2004) A systems – theoretic approach to safety engineering: a case study MIT ESD external symposium, Cambridge, MA
Leveson N (2011) Engineering a safer world: systems thinking applied to safety. https://mitpress.mit.edu/sites/default/files/titles/free_download/9780262016629_Engineering_a_Safer_World.pdf
Ogata K (1997) Modern control engineering. Pearson US Imports
Perrow C (1984, 1999) Normal accidents: living with high-risk technologies. Princeton University Press, Princeton
Sterman JD (2000) Business dynamics: systems thinking and modeling for a complex world. Irwin McGraw-Hill, Boston
Stringfellow Herring M, Owens BD, Leveson N, Ingham M, Weiss KA (2007) A safety-driven, model-based system engineering methodology, Part I. MIT Technical Report, December 2007, 56p

Chapter 6
Using Stamp in the Risk Analysis of a Contaminated Sediment Treatment Process

Karim Hardy and Franck Guarnieri

Processes for remediation (removal of pollution or contaminants) of contaminated sediments have become very efficient. These technologies, which are particularly complex, call for a comprehensive approach to risk analysis which characterizes all threats (to humans, equipment, local residents, the environment etc.). The STAMP accident model (Systems-Theoretic Accident Model and Processes) is an example of such a comprehensive approach, and it has been chosen to characterize the risks associated with Novosol®, an innovative remediation process. Risk analysis is carried out through the application of STPA (STamP-based Analysis).

This chapter is organized into three sections. The first describes the industrial context for treating contaminated sediments. The second introduces the Novosol® process. Finally, the third section describes the concrete application of the STPA technique to the process.

6.1 The Industrial Context

This section presents the industrial context in which the STPA technique is applied.

K. Hardy
College of Aeronautics, Worldwide Campus, Embry-Riddle, Aeronautical University, Daytona Beach, FL, USA
e-mail: karim.hardy@erau.edu

F. Guarnieri (✉)
MINES ParisTech/PSL Research University, CRC, Sophia Antipolis Cedex, France
franck.guarnieri@mines-paristech.fr

F. Guarnieri, E. Garbolino (eds.), *Safety Dynamics*, Advanced Sciences and Technologies for Security Applications,
https://doi.org/10.1007/978-3-319-96259-7_6

6.1.1 The Problem of Contaminated Sediments

The natural environment is frequently subjected to a wide variety of industrial, urban and agricultural wastes that are rich in pollutants and represent major sources of contamination. Typically, these contaminants remain in suspension in silt during sedimentation and are and are generally neutralised as a result, with no real danger. However, if the physicochemical condition of the contaminated sediment changes, the contaminant may become dangerous to the environment or human health. The most common example is the heavy metals that permeate rocks: the concentration levels observed today are higher by a factor of hundreds or thousands than they were at the start of the industrial age.

The damage caused by contaminated sediments has major environmental, social and economic costs. Sediments can increase the need for dredging because they increase the risk of flooding and reduce the draft of navigable rivers. Also, contaminated sediments cause health problems for ecosystems and nearby populations, and result in substantial losses of income due to contamination and reduced numbers of animal and plant species.

The main contaminants found in sediments (cadmium, copper, chromium, lead, zinc, polychlorinated biphenyls (PCBs), polycyclic aromatic hydrocarbons (PAHs) and arsenic) originate from industrial activity (Table 6.1). This contamination can have very different, but dramatic, health and biological effects (such as genetic or ecophysiological damage and changes in species behaviour).

This section first presents the approach to the management of contaminated sediments and then goes on to examine the various contaminated sediment treatment technologies.

6.1.1.1 Approach to the Management of Contaminated Sediments

Contaminated sediments can pose serious problems for both human health and the environment (Hardy and Guarnieri 2011a, b); they are a major source of chemical pollutants and are dangerous to human populations and aquatic organisms. Human exposure comes from direct contact, by eating fish or shellfish that have ingested toxic chemicals, or from drinking water that has been exposed to contaminated sediments. Aquatic organisms, especially benthic species[1] are continually exposed to the contaminated sediments, which induces both acute and chronic toxicities. The accumulation of toxic chemicals in aquatic organisms can render them unfit for human consumption and pose a threat to wildlife higher up the food chain. To preserve the environment, contaminants in sediments must be treated to avoid or limit the potential exposure of human populations and aquatic organisms. In addition, the degradation of the quality of the sediment has to be taken into account.

[1] The adjective benthic means relating to the bottom of a body of water (sea or freshwater), regardless of depth.

Table 6.1 Sources of sediment contaminants (Source: European Commission)

Industrial Sector	Cadmium	Copper	Chromium	Lead	Zinc	PCB
Steel/iron				●	●	●
Aluminium	●		●			
Anti-fouling paint		●		●		
Electrical appliances	●	●		●	●	●
Automobile	●	●	●		●	●
Batteries				●		
Rubber					●	
Shipyards	●	●	●		●	●
Chemical	●		●			
Leather/tanning			●			
Detergents/surfactants					●	
Water/gas/electricity distribution					●	
Explosives		●				
Extraction of precious minerals				●	●	
Oxide production		●	●		●	
Metal finishing	●	●	●	●	●	
Steam power	●	●	●	●		
Electroplating		●	●	●	●	
Munitions		●			●	
Photography			●			
Pigments/inks				●		
Printing plates					●	
Plastics				●		
Metallurgical processes					●	
Oil refining				●		
Diverse sources	●	●	●	●	●	
Wastewater treatment		●	●	●	●	●

Techniques for the remediation of contaminants in sediments range from natural solutions (burial), to elimination, or completely isolating them from the environment. When contaminated sediments are found in areas that must be dredged to keep waterways open, removing the sediment is the only practical solution, but the subsequent treatment of the removed material becomes a problem. Most countries that process contaminated sediments follow the procedure adopted by the United States' Environmental Protection Agency (EPA), which is shown in Fig. 6.1. The five stages of this process are described below.

Step 1—Detection of the Sources of Contamination

Sediment contamination is a process that is influenced by many variables, including: the source of contamination; the types of contaminants; the sedimentary and hydrological environment; the size, composition and distribution of sediment particles; and the presence or absence of aquatic life. The probability that contamination will be a problem at a particular site must be estimated based on the site-specific

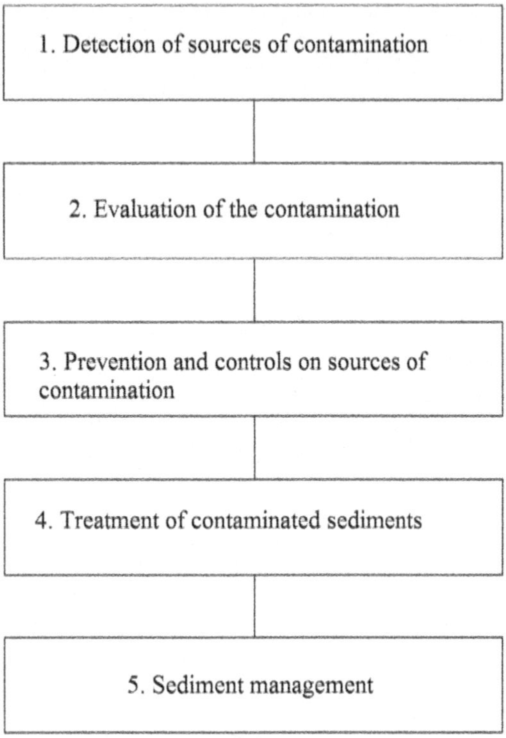

Fig. 6.1 General approach to the remediation of contaminated sediments. (US-EPA 2005)

information available about these variables. This information comes from a range of sources, such as rehabilitation monitoring programmes, previous *in situ* studies, records of earlier drainage activities, fishing and geographical zoning. Sediment contamination is very often associated with very poor water quality: the capacity of sediments to retain contaminants means that the levels of contamination exceed safe limits set by water quality standards.

Step 2–Assessment of the Contamination
When a preliminary analysis indicates a possible contamination problem, the sediment has to be characterised as completely as possible. This includes assessing the environmental threat. When choosing a sediment assessment method, protection of worker should be the priority. Assessment methods vary according to what contaminant is being assessed and whether the expected negative effects are on the aquatic environment and its wildlife or on human health.

The simplest assessment directly compares observed sediment contaminant concentrations with predetermined criteria. Sediment quality criteria are more difficult to implement than criteria related to water quality because of the number and complexity of the factors that can influence the biological effects arising from particular levels of concentration of a given chemical product. These adverse effects of contaminated sediments include:

- the type and extent of all identifiable effects on ecosystem health and well-being, and not just the effects on the plankton, fish, shellfish or wildlife that are exposed;

- the concentration and distribution of pollutants found in biological, physical and chemical processes in the ecosystem;
- the effects of pollutants on biodiversity.

Step 3–Prevention and Control of the Sources of Contamination
It is important to determine the extent and origin of any newly-discovered contamination. The response partly depends on the extent of the contamination: if a contaminant is identified in several places, it is clearly inefficient to only manage one site. Contaminated sediment management requires a comprehensive approach. Sources of contamination are generally so numerous that an isolated action (e.g., on a single discharge) is often ineffective; consequently, a comprehensive and systemic treatment programme is needed, together with accompanying specific regulations. The presence of a contaminant at several geographical sites does not necessarily indicate that large-scale control action is needed. In cases where contamination is limited to a single site, however, local control may be sufficient, and large-scale tests may be inadequate or ineffective.

Although there are standardised approaches to the treatment of contaminants at national, regional and local levels, each contaminated site has a unique configuration and composition. It is therefore usually necessary to adopt a site-specific approach, especially for particularly sensitive locations. Until legislation can be implemented, immediate control measures need to be put in place as any delay could have an adverse (possibly irreversible) effect on the environment.

Three types of sources of contamination can be distinguished:

- localised sources, such as industrial or municipal discharge;
- non-localised sources: agricultural spraying; floodwater; atmospheric contaminants, etc.;
- other sources: spills, drainage systems, etc.

If the contamination is limited to a single site, the first factor to consider is the age of the contamination; for example, if the contamination source is historical then it is too late to use source control.

When the source of the contaminant has been brought under control, accumulated sediment contaminants may still persist. In these cases, recovery is preferred to a source control approach.

The best way to identify the link between sediment contamination and its potential source is to take a contaminant-by-contaminant approach. Toxicity tests cannot directly identify particular contaminants, although they can play a key role in an initial sediment analysis, and toxicity identification techniques can be used to determine which contaminants are responsible for the observed toxicity. The geographical relationship between a peak concentration of a contaminant in a sediment and its potential sources can also be used to link sediment contamination with a potential source.

When there are multiple sources of contamination for a given contaminant, additional evidence should be sought to link that contaminant to its source. Contaminant concentration levels can be observed in the surface sediments, and the contamination reaches its peak at the site of the spill. As the level of contamination can vary as a function of the depth of the reading in the sediment record, the history of the contamination of the sediment can be traced. In this way it becomes possible to establish a site's fingerprint through the analysis of the occurrence or concentration of one or more contaminants (Fig. 6.2).

Step 4–Treatment of Contaminated Sediments

The environmental benefits of contaminated sediment treatment need to be weighed against the costs of the treatment and its potential impact on the environment. For example, it is likely that a benthic habitat will be destroyed through sediment dredging, or that sediment contaminants will be dispersed by drainage operations.

The minimum goal of sediment treatment is to achieve an acceptable environmental quality, without altering natural processes, over a reasonable period of time. Normally, sediment that either fails to meet the relevant quality criteria, or has adverse environmental impacts must be considered with a level of risk as unacceptable.

In all cases, several questions have to be addressed when deciding whether or not to treat contaminated sediments (US EPA 2005):

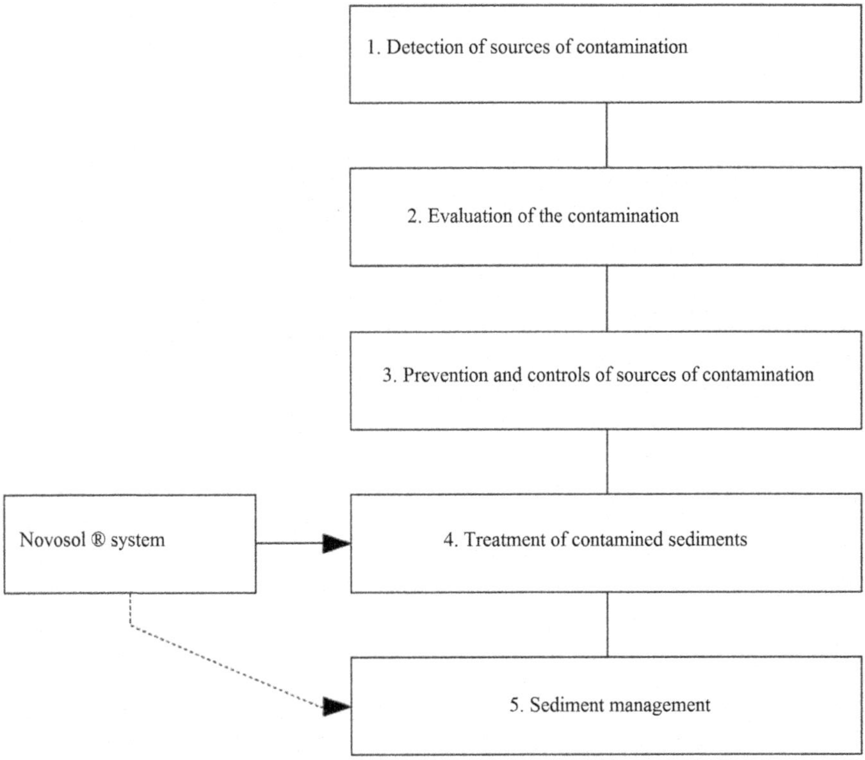

Fig. 6.2 The role of the Novosol® process in the overall approach to contaminated sediment management

- Can the contaminated sediment be processed in a reasonable amount of time?
- Will the treated sediments disperse naturally or by another mechanism?
- Does the contaminated sediment pose a danger to remote areas?
- Do technically and economically feasible alternatives exist for the treatment of the contaminated sediments?
- Are the positive long-term effects of the treatment beneficial to the environment compared with any short-term adverse effects?

The responses to these questions generally point to three possible courses of action (Fig. 6.3):

- Take no action and allow natural sedimentation to take its course;
- Treat the contaminated sediment *in situ*; or
- Dredge contaminated sediments and send them to another site for treatment or storage.

These three options can be combined; for example, a decision may be made to dredge sediments in one area of a site and to treat them in another. The choice depends on: the expected environmental benefits; the potential adverse impacts on health and the environment; the natural decontamination time; and the feasibility and costs of different treatments, transport and storage in landfill.

With respect to the different courses of action, if the principal aim of the remediation process is to limit health and environmental risks, sediments that are left *in situ* are probably more contaminated than those that are drained. Therefore, different levels of treatment may be required before the sediment can be remediated. In some cases, it is possible to separate sediments and treatment residue into contaminated and uncontaminated fractions. This separation is common to several treatment technologies, including those that remove contaminants by *calcination* (such as the

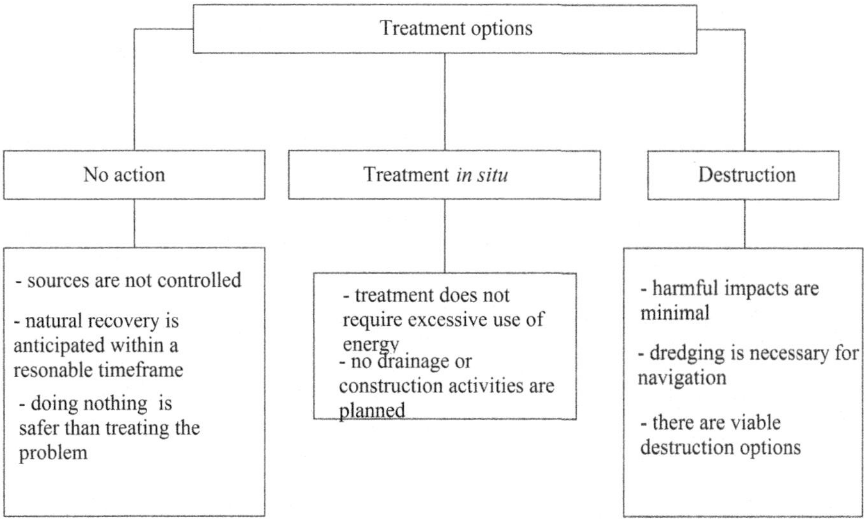

Fig. 6.3 Treatment options for contaminated sediment. (US-EPA 2005)

Novosol® process). The contaminated portion of the sediment can be stored in a protected area, while the treated, decontaminated sediment can be reused. To reduce costs, however, contaminated sediment treatment technologies must be able to reduce the chemical concentrations to acceptable levels.

The methods available for treating contaminated sediment fall into five general categories:

- Reducing the volume of contaminated sediments;
- Neutralisation of the contaminant;
- Destruction of the contaminant;
- Extraction of the contaminant;
- Thermal separation.

Step 5–Sediment Management

The treatment step is immediately followed by a sediment management step, which consists of four phases:

1. The aquatic discharge of decontaminated or uncontaminated sediment;
2. The reuse of sediment, in dam construction, for example, or in the case of the Novosol® process, in road embankments made of treated or decontaminated sediment;
3. Aquatic overspill that confines contaminated sediments through recovery in an aquatic site, or storage in a watertight location;
4. The construction of a facility dedicated to highly-contaminated sediments, where treatment can lead to the emission of atmospheric waste.

This contaminated sediment management process combines several processing techniques, which are presented below.

6.1.1.2 Contaminated Sediment Treatment Technologies

The treatment of contaminated sediments raises increasingly difficult technological, economic and environmental challenges. Treatment can be used to achieve acceptable pollutant levels so that contaminated sediments may eventually be reused. The results of the sediment characterisation phase are used to help to select the appropriate treatment technology and, above all, to estimate the cost. Certain treatment techniques may have additional environmental impact, due to the discharge of water or gas.

Table 5.1 shows most of the classical contaminated sediment treatment technologies. None of these techniques can remove, store or treat contaminated sediment without perturbations and without the release of contaminants. The changes applied to the sedimentary system at the bottom of an aquatic site can cause contaminant suspension in the water column, which the treatment process must reduce to the lowest possible level.

The choice of treatment is biased towards those that can be applied *in situ*, because this reduces the risks associated with transportation, as well as financial costs. *In situ* treatment does, however, require an extensive amount of space, which is rarely available.

In most cases, sediment is dredged and then stored or treated off site, mainly because the high social acceptance of this approach outweighs the higher costs of extraction, transportation and processing. The removal and transport of contaminated sediments are delicate steps that can introduce contaminants into a previously clean site. The decision to dredge depends on the nature of the sediment, the depth of the aquatic site, the thickness and volume of the sediment, the distance to the treatment site and the availability of equipment.

The method of transporting the dredged products depends on the distance between the dredging and treatment sites, and is usually by pipeline, barge, rail or truck. The selection of the type of transport also depends on the type of dredging technique used—mechanical, hydraulic or pneumatic—and the choice of treatment of contaminated sediments.

After dredging and transport, the sediment undergoes a pre-treatment phase, which includes the removal of surplus water and the filtering of particles to eliminate the larger particles. This pre-treatment phase has four advantages, which are related to the ability of contaminants to bind to fine-grained particles;

- It reduces the water content to facilitate transport to a storage or treatment site;
- It reduces the volume of sediment to be treated or placed in confinement sites;
- It fosters or accelerates the sedimentation of solids and separates recyclable materials from those to be treated or stockpiled;
- Finally, it makes it possible to separate materials into different categories so they can be appropriately treated.

The elimination of water contained in the sediment brings it closer to a solid state for handling and transport, and prepares it for further processing. Eliminating excess water is generally efficient and inexpensive, but requires time and a large amount of space. It is usually performed using centrifugation, filtration or gravity. Chemicals such as flocculants can be added to accelerate the deposition of suspended solids. Filtering separates sedimentary particles based on their physical properties (size, density, mass, etc.): the coarse fraction can be reused, after analysis, as backfill or as building material. This can substantially reduce the costs of decontamination.

The fundamental objective of all contaminated sediment treatments is to extract, immobilise, remove or neutralise contaminants. Different processing techniques have different advantages and disadvantages, mainly due to the limited number of contaminants that they can treat (some techniques only treat one type of contaminant). The choice of treatment method is influenced by: the propensity of contaminants to leaching[2]; chemical stability of the contaminants, the characteristics of the sediment and its contaminants; sediment homogeneity; and the social and health impacts. Some of the three major classes of treatments (biological, physicochemical, and stabilisation, see Table 6.2) are used in the Novosol® process, described below.

[2]According to Larousse "an operation that consists of slowly passing a solvent through a product that is suitably pulverised and deposited in a thick layer to extract one or more soluble components" or "chemical dissolution of certain components of a material used to extract ore, metals, or valuable minerals".

Table 6.2 Treatment techniques for contaminated sediments (US-EPA 2005)

Technique	Application	Characteristics	Efficiency	Cost
Biological				
Biolixiviation	Pesticides, hydrocarbons, PCBs, chlorinated aromatics	pH 4.5–8.5 59–167 °C 40–80% hydration	Depends on the treatment volume	Moderately high
Physicochemical				
Dechlorination	Dioxins, PCBs, chlorobenzene	pH > 2 158–302 °C < 20% hydration	Efficiency >98% for PCBs	High
Solvent extraction	PCBs, volatile organic compounds, aromatics, metals	Organic compounds <40% Solid portion <20%	Efficiency ~ 90% for PCBs	High
Soil leaching	Heavy metals, aromatics, PCBs, pesticides	Particle size 0.063–2 mm	90–99% of volatiles and 40–90% for semi-volatiles	High
Solidification/ stabilisation	Inorganic compounds, oily sludge and solvents		100% effective on inorganic compounds	Moderate
Thermal				
Calcination	Volatile and semi-volatile compounds, dioxins	< 50% hydration particle size 1–2 mm	> 99% for organic compounds	Very high
Desorption at low temperature	Volatile and semi-volatile compounds		99%	High

6.2 Novosol®

Solvay SA's Novosol® process (Novosol® 2010) combines several techniques for treating contaminated sediments to stabilise mineral residues that have been contaminated by heavy metals or organic compounds. The process constitutes a system as defined by von Bertalanffy (1968).

Novosol® includes several innovative technologies that address problems associated with decontamination. They include the Neutrec process,[3] also developed by Solvay SA, which is used to purify the smoke emitted during the Novosol® process.

[3] The Neutrec (www.solvay.com) process is "based on the dry injection of finely-ground sodium bicarbonate in the flue gas to be purified. The sodium bicarbonate neutralises the acids (hydrochloric acid, sulphur dioxide, hydrofluoric acid, etc.) with a very high efficiency. Through the combined injection of activated carbon or lignite coke, fumes are cleaned of heavy metals and dioxins/furans, and meet the most demanding legislator standards. The process is totally dry; no aqueous effluent is generated or treated".

The Novosol® process is outlined below. After providing a background on how the process was developed. The technical aspects are described, focusing on the physicochemical treatment process. Then, the socio-technical aspects of the system are presented, including details of the various actors who are involved in both the development and operational phases of the process.

6.2.1 Background

The Novosol® process is the culmination of 15 years of research into the stabilisation of mineral residues contaminated by heavy metals and organic compounds. Solvay SA began developing the process in 1993 (Depelsenaire 2006), initially to treat incinerator ash. Beginning in 1999, it was subsequently extended to treat a wider range of contaminated sediments. The work began as a partnership with the Free University of Brussels (Breugelmans 2007) and the process was formulated in response to a need in the Belgian Walloon region to treat sediments contaminated by heavy metals. This process comprises two treatment steps (Novosol® 2010): phosphating, followed (optionally) by calcination. After promising initial results and patenting,[4] work continued at Solvay's research centre.

In 2004, conclusive results led to the implementation of an industrial pilot project in collaboration with institutional and industrial partners. This was followed a year later by a second industrial pilot project at one of the Solvay group's facilities. At the end of 2005, a 100 metre test road was built, re-using 30% of the treated river sediments, in partnership with a road construction company. In addition, around 25% of the sediment that was treated with the Novosol® process was incorporated into the production of bricks. The foundations of a comprehensive production facility were completed in late 2006 and an industrial facility was constructed in 2011 by Solvay SA's partners.

Solvay only grants licenses and does not use or operate any Novosol® installations itself. In 2010, limited licences for the use of the Novosol® process were granted to two operating companies, SEDISOL[5] and SIFA. Both companies are licensed to use the Novosol® technology to treat contaminated sediment. At the same time, both companies are in an industrial partnership with a company responsible for Novosol® development: SEDISOL works with SERIP,[6] while SIFA works with Studio Altieri.[7]

The Novosol® process addresses the problem of treating mineral residues containing heavy metals and organic matter. These residues include not only polluted sediments from canals and ports, but also the residue from activities that are man-

[4] Patents: in Europe EP 0883585 B1 (12 January 2000) and EP 0899000 B1 (3 November 2004). In the United States 6,132,355 (17 October 2000) and 6,254,846 (3 July 2001).

[5] http://www.sedisol.be/

[6] http://serip-france.pagesperso-orange.fr/

[7] http://www.studioaltieri.it/

aged by national and local authorities, such as ash from waste incinerators. Contaminated residues are also produced by private companies, and include car crushing waste and industrial sludge. These areas of activity can generate many different types of contaminated waste (see Table 6.1). Disposal may go beyond simple environmental protection and be an absolute necessity.

6.2.2 *The Physicochemical Process*

6.2.2.1 Step One: Phosphating

The first step of the Novosol® process is phosphating (Figs. 6.4 and 6.5). Phosphoric acid is added to the contaminated sediments to ensure the stabilisation and reduction of the solubility of heavy metals. From an operational point of view, the phosphating unit is designed to treat 100,000 m^3/ year of raw sediments at 50% humidity for 200 days/year (five days/week). To achieve this, it treats 100 m^3/hour for five hours/day and only requires the presence of operators during the day.

The phosphating phase comprises several steps. Under Solvay engineering specifications, each step is the responsibility of one of three operators (designated as A, B and C) and a team leader (Figs. 6.4 and 6.5). Each operator at a Novosol® site has one or more areas of responsibility that are linked to their operational tasks.

Operator A (Figs. 6.4 and 6.5) is responsible for:

- filtering;
- storage;
- pumping and phosphating sediments;
- measuring the density of sediments;
- gas treatment.

Operator B is responsible for:

- the management of phosphate sediments;
- liquid storage basins.

Operator C is responsible for:

- utilities;
- truck loading;
- supply and shipping logistics.

The process is mainly used for the treatment of dredged mud and is part of a comprehensive approach to the management of contaminated sediment (see Fig. 6.2). Siltation of waterways and ports can become a major obstacle to fluvial navigation. The sediments are often loaded with pollutants from urban, industrial

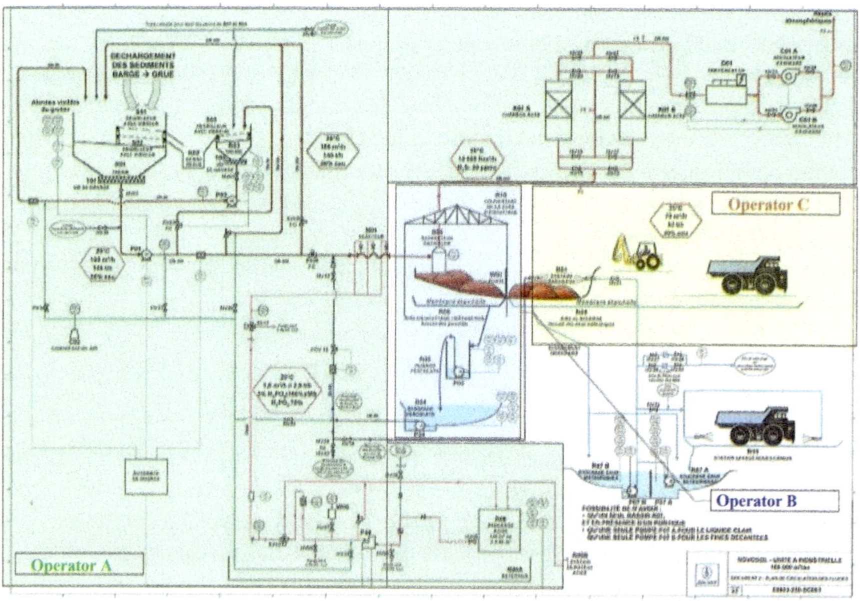

Fig. 6.4 Novosol® flowchart showing the responsibilities of each operator (A, B and C). (Source: www.solvay.com)

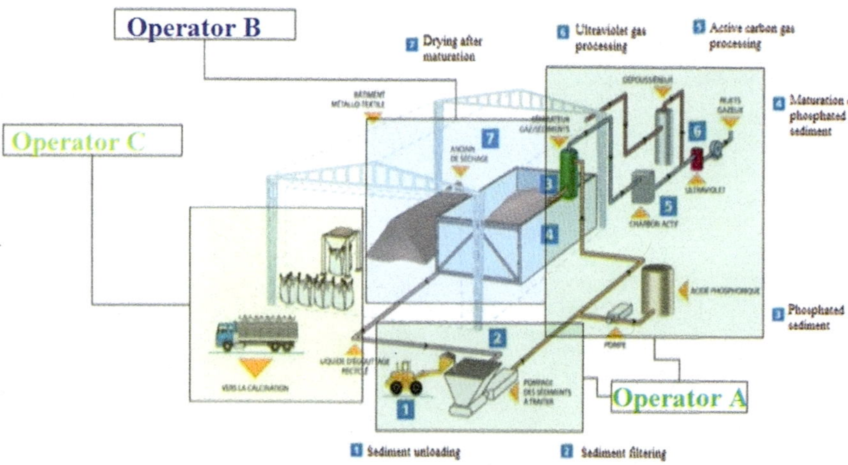

Fig. 6.5 Phosphating stage in the Novosol® process showing the responsibilities of each operator. (Source: Solvay.com)

and agricultural waste, and can increase the risk of flooding and pollute drinking water. The most common solution is to dredge the affected area to extract the sediment.

Heavy metals are treated by adding phosphoric acid to form calcium phosphate. After phosphatising, the material is stored for 24 h on a geotextile drainer. It is then poured onto drying windrows, which reduces the water content by 40% and immobilises the heavy metals. The process includes a gas cleaning system based on active carbon filtration and washing.

Phosphating is optionally followed by a calcination phase, depending on the particular treatment needs.

6.2.2.2 Step Two: Calcination

The phosphating phase produces organic material that can be removed by calcination (Fig. 6.6). Calcination increases the stability of heavy metals that were separated out separated out by phosphating, and generates recyclable products. It is during this phase that the Neutrec process is used to purify the fumes from the calcination of the sediment.

Several pilot Novosol® units were built to validate the performance of the process and evaluate investment and operating costs (Hardy and Guarnieri 2011a, b). These test units confirmed that treated residues could be recycled (in dyke and road construction, brick-making, etc.), instead of sending them to landfill. These recycling methods were complemented by the development of cement based on residues from the process.

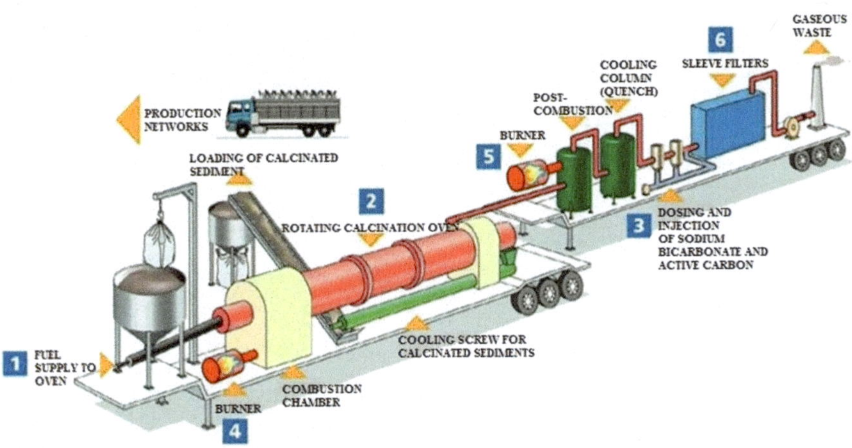

Fig. 6.6 Calcination in the Novosol® process. (Source: Solvay.com)

6.2.2.3 Dynamic Model of Phosphating

The following analysis focuses on phosphating alone. It is based on the industrial requirements defined by Solvay SA. The calcination phase is not discussed.

From a strictly technical point of view, the dynamics of the Novosol® process are simple (Fig. 6.7). As the process is quantifiable, it can be used to study operational and behaviour scenarios. Here, all interactions, including those related to human and organisational factors, are included in the analysis.

The analysis is based on a model that represents the dynamics of the Novosol® process. The aim is to achieve a level of performance that is consistent with system safety approaches (Fig. 6.8). This dynamic model includes all of the variables that are defined during the application of the STAMP model, which makes it easier to understand all the interactions in the system (Table 6.3). The set of defined variables forms part of an approach to performance improvement based on risk management and system safety.

The overall structure of the Novosol® system is shown in Fig. 6.9 as a static model that highlights hierarchical levels and the interactions between them. This schema follows the procedures outlined in Step 2 of the STPA method in the static phase and takes into account all of the social actors involved in the process. This structure is used to help define the roles and responsibilities (controls) that operate between levels.

This static model provides the basis for the dynamic model. It includes all of the variables that are retained when defining the problem and consequently allows the development of a model as a function of the particular objective: in this case, a safety assessment that aims to optimise the performance of the Novosol® system.

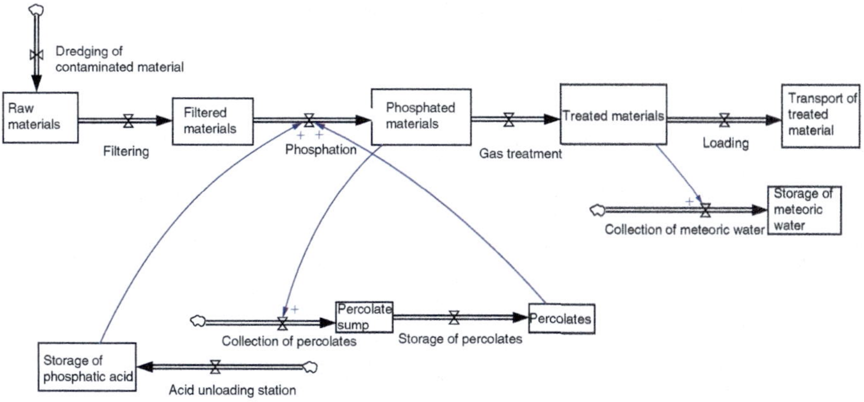

Fig. 6.7 Dynamic model of phosphating in the Novosol® process

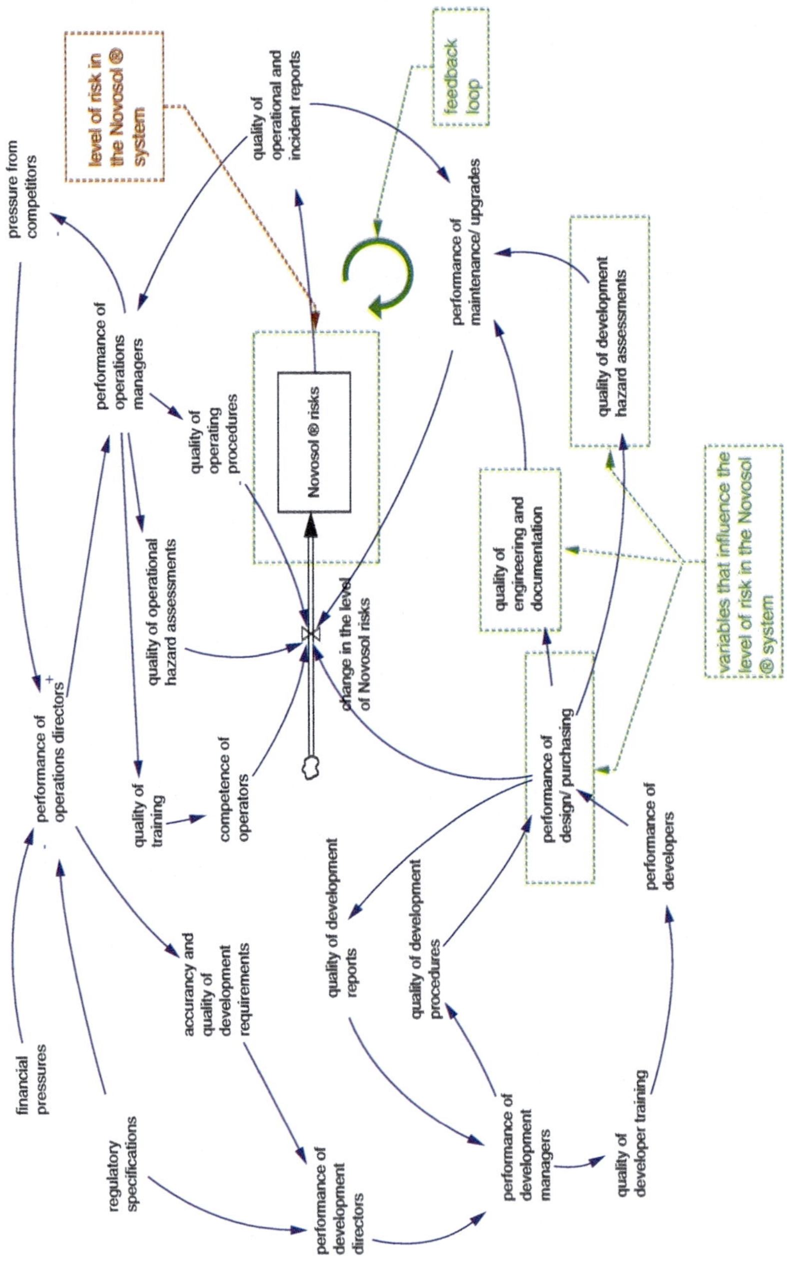

Fig. 6.8 Influences in the Novosol® system during development and operation, focused on the level of risk

Table 6.3 Structural variables used in the Novosol® system

Structural variables in the Novosol® system	
Related to a hierarchical level of performance	Related to a level of quality (controls)
Operations managers	Regulation
Development managers	Development requirements
Management of development	Development reports
Developers	Documentation and engineering
Design and purchasing	Hazard assessment in the design phase
Maintenance and upgrades	Operating and incident reports
Management of operations	Operating procedures
	Evaluation of operational hazards
	Training of operators
	Development procedures
	Developer education

6.3 The STPA Hazard Analysis

Prior to the analysis, Solvay SA defined the following safety requirements:

- the application of a risk management approach to eliminate risks in the Novosol® process in order to achieve an acceptable level of risk;
- the formulation of recommendations to reduce the level of risk in order to maximise safety, taking into consideration performance, cost and time requirements.

The approach that was adopted involved establishing integrated, global safety requirements using the STPA technique to improve the level of safety throughout the system's lifecycle.[8] The Novosol® system includes potentially inadequate controls (failure to evaluate hazards, untrained operators, unclear or undefined operating procedures, etc.) at many hierarchical levels, which can affect the overall level of safety. The five steps of the STPA hazard analysis technique (Fig. 6.10) were applied to the Novosol® system, as described below.

Step 1–Preliminary Analysis of System Risks and the Definition of Requirements and Constraints

In a safety assessment, a preliminary analysis of the risks and hazards is performed at a system level in order to define the safety requirements and constraints to be integrated. This analysis should take place as early as possible in the life cycle of the system.

The preliminary analysis of risk in the Novosol® system was performed at two levels: the physicochemical process; and the socio-technical system. This decision

[8] All the reports and the results of the risk analyses for the Novosol® destined to be used at Solvay's industrial facilities were provided as paper-based report for the HAZOPS analysis and a DVD containing the HAZOPS analysis together with recommendations and the evaluation of occupational risk. This material can be edited and updated throughout the lifecycle of the Novosol® system.

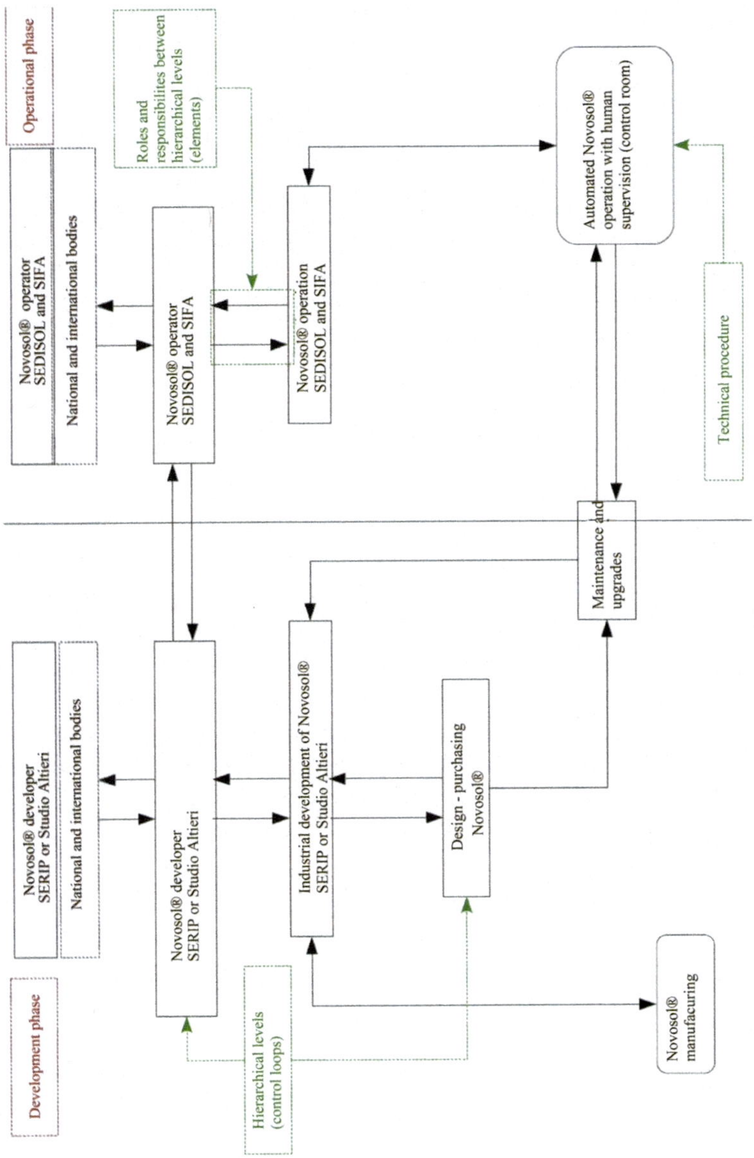

Fig. 6.9 Overall description of the structure of the Novosol® system

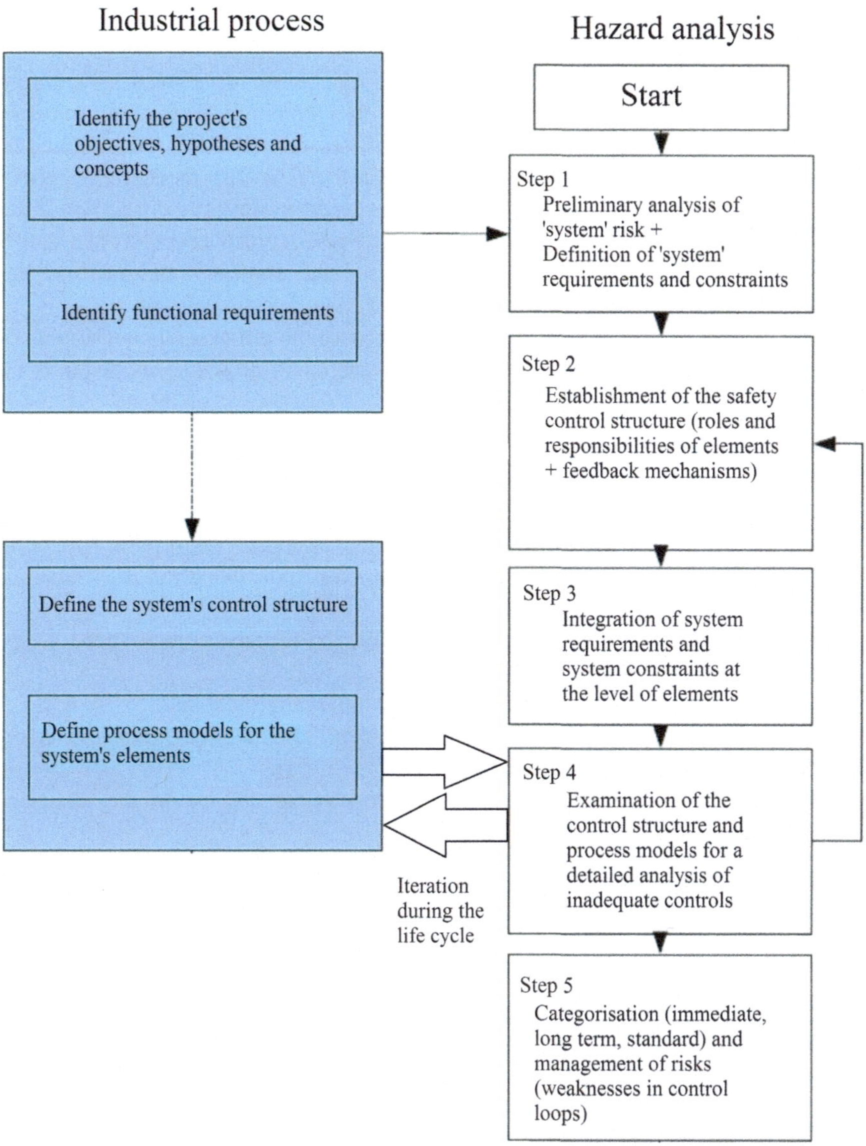

Fig. 6.10 STAMP safety assessment (STPA). (Source: adapted from Leveson et al. 2004)

was consistent with the requirements of the Solvay group and the STPA hazard analysis methodology.

The initial analysis was designed to provide a response to the needs of industry in terms of risk management of phosphating within the Novosol® system. An occupational risk assessment and HAZOPS analysis[9] (Ericson II 2005) were applied to the Novosol® process. Like the STPA technique, the HAZOPS methodology seeks to identify any discrepancies between the desired system state and the current state. However, unlike the STPA technique (which addresses control problems in a socio-technical system, taking into account human and organisational factors) the HAZOPS analysis focuses on technical parameters in a technical system.

This HAZOPS approach was used to characterise the initial safety constraints in the phosphating phase, with a focus on process engineering. It was divided into the following five steps (Fig. 6.11):

- the division of operational units as a function of the areas of responsibility for each operator (see Figs. 6.4 and 6.5). Operator A is responsible for filtering, storage, pumping and phosphating sediments, measurement of sediment density, and gas treatment; Operator B is responsible for the management of phosphate sediment and the liquid storage basins. Operator C is responsible for utilities, loading trucks, and supply and shipping logistics;
- the division of each operational unit into operational time periods (pre-startup, startup, operations and shutdown);
- the division of each operational time period into operational tasks;
- an analysis of each operational task and the identification of deviations, based on a HAZOPS grid;
- the definition of recommendations and constraints applied to the technical Novosol® process.

Table 6.4 shows an excerpt from the HAZOPS analysis.

The results of these analyses[10] were used to formulate a number of safety recommendations that were used to improve both the design of future Novosol® installations and the associated level of safety prior to commissioning.

Although the HAZOPS analyses met Solvay's industrial requirements, the approach is unlike the STPA technique, which has never been used in the context of a system such as Novosol®. The technique was therefore introduced to extend the analysis to the entire socio-technical system. This analysis included consideration of all of the human and organisational factors related to the site, and the companies that developed and operated the system.

For each hierarchical level of the system, system requirements and constraints were defined. For the Novosol® operator (Solvay during the development stages, and currently SEDISOL and SIFA), the requirements and constraints identified using the STPA method are shown in Table 6.5.

[9] A Hazard and Operability analysis (HAZOPs) (Ericson II 2005) is a hazard analysis technique that seeks to identify deviations in a system or process, usually physical or chemical.

[10] These analyses were delivered to Solvay in 2009.

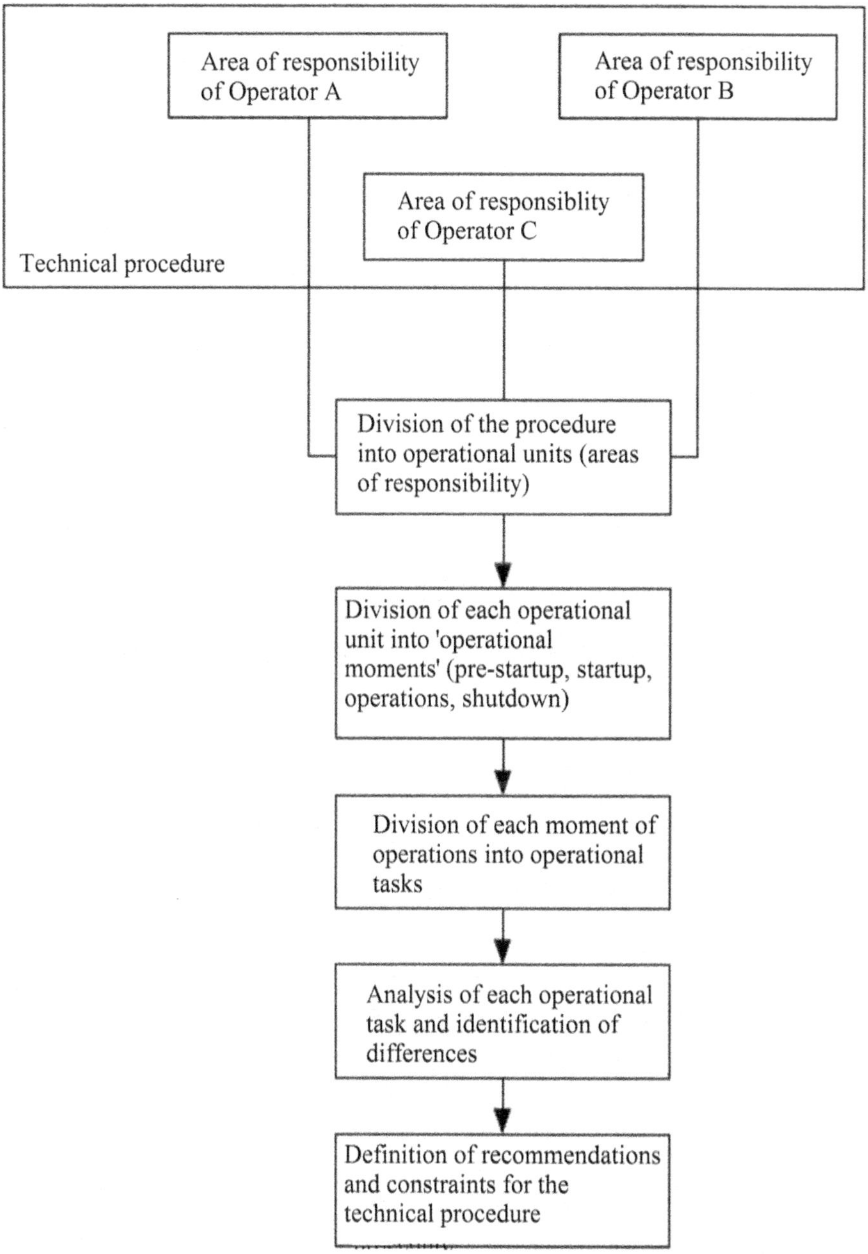

Fig. 6.11 HAZOPS approach to the technical procedure

Table 6.4 Extract from the HAZOPS analysis of the technical procedure showing recommendations and safety constraints

Unit	Operator A – pre-startup
Task	2 – Control of the water content of sediment
Element	Raw sediments
Deviation	Inadequate control of the water content of sediment
Parameter	Dryness
Keyword	Lower / none
	Too much
Result	Excessive water content
	Suboptimal treatment of raw sediments during phosphating
	Insufficient water content
Cause	Too much water
	Too little water
Risk	3B
	3B
Constraints	Systematic check of dryness for each new barge
	Add water at R01 and R03 hoppers to reduce dryness

Once all of the requirements and constraints for all hierarchical levels had been defined, the hierarchical control structure was established.

Step 2—Establishment of the Safety Control Structure

The safety control structure describes the whole of the Novosol® system from a static viewpoint. It is constructed by linking the different hierarchical levels using the interactions between the elements. This step is based on Step 1 and includes all of the stakeholders who are defined by the requirements and constraints of the system. Each level or element of the control structure has roles and responsibilities that are designed to ensure that system safety constraints are applied. By incorporating the roles and responsibilities, it becomes easier to determine the links between elements (Fig. 6.12).

The aim of the structure is to characterise the controls between elements. It sets the limits for the analysis and then determines potentially inadequate controls between levels. The roles and responsibilities are used in Step 3 of the analysis to support the definition and integration of constraints at the level of each actor in the structure.

Step 3—Integration of System Requirements and Constraints

In this step the system requirements and constraints defined in Step 1 are incorporated at each appropriate hierarchical level of the safety control structure established in Step 2. This is done from the point of view of the interactions between elements. This step makes it possible to define system requirements that are then translated into safety requirements from the point of view of the interactions between the elements in the safety control structure.

Table 6.5 Example of the definition of requirements and constraints for the controller: the operating company

Novosol® operator (SEDISOL or SIFA)
Constraints and safety requirements • Treat sediments contaminated by organic compounds and heavy metals. • Responsible for inspections and the creation of reports on the operation and development of Novosol® in conjunction with national and international bodies. • Responsible for defining the requirements and the operating performance of the Novosol® process under national and international regulations. • Ensure that operational requirements are properly integrated and transmit any event to the next level in the hierarchy with respect to reporting (operation and accidents).
Context in which decisions are made • Ecology as a societal problem. • Sediment pollution is a new market in terms of the requirements of local communities and populations. • Processes that are under development and prototyping require continuous adaptation to achieve the required performance. • The company is carrying out research or is in partnership with industrialists who may be potentially interested in Novosol® technology.
(potential) inadequate control actions • Unidentified hazards. • Inadequate coordination with the company in charge of development. • Insufficient attention given to accident/ incident reports. • Insufficient attention given to operational reports. • Insufficient or improper integration of national and international regulations. • Development requirements incorrectly defined with respect to the operating company's strategy.
Defects in the cognitive model • Thinking that the final Novosol® process is totally safe. • Not being aware of all the risks during the operation of the process.

A higher hierarchical level, for example, the Novosol® developer (SERIP or Studio Altieri), defines requirements for the lower hierarchical level (Novosol® operator). The lower level (SEDISOL or SIFA) must provide feedback (control confirmation) by providing reports to the higher level. This sequence applies to each interaction and each variable. Specifically, at this level of the structure, SERIP or Studio Altieri must define and implement the requirements for a Novosol® installation or entity responsible for industrial operations. In return, and in order that the management of SERIP and Studio Altieri are informed of the successful integration of these requirements (controls), the service or unit must provide reports describing the progress of the project or any difficulties that may be encountered.

Therefore, for the hierarchical level corresponding to the Novosol® developer, the following wording could be used, "The company responsible for Novosol® development (SERIP or Studio Altieri) must define requirements for the Novosol® operator (SEDISOL or SIFA). In return, the Novosol® operator (SEDISOL or SIFA) must provide progress reports to the Novosol® developer (SERIP or Studio Altieri)."

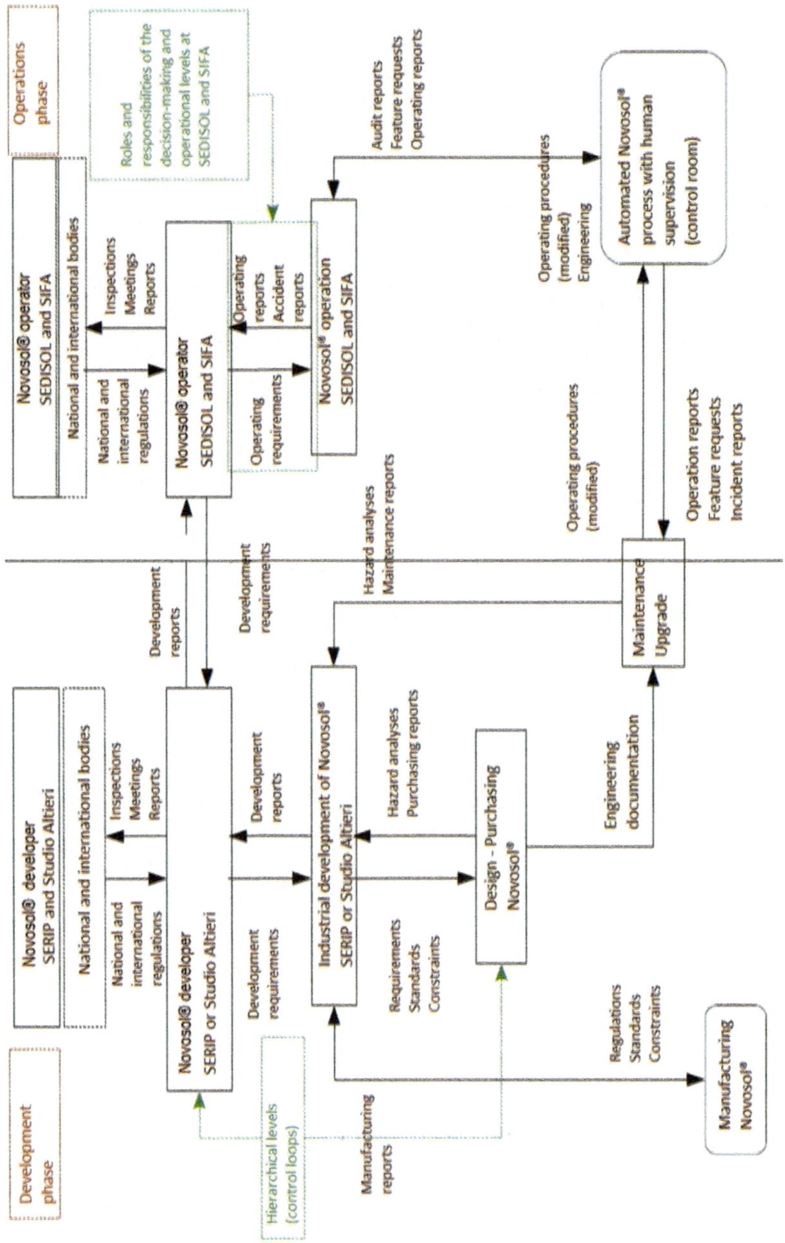

Fig. 6.12 Structure of the Novosol® system following the application of the STPA analysis technique

Step 4—Detailed Analysis of the Control Structure and Process Models to Identify Inadequate Controls

In this step, a detailed analysis is carried out to identify any potentially inadequate controls that may contribute to the occurrence of an accident. The analysis is based on four types of inadequate controls:

- a potential control action is not carried out;
- a potentially incorrect or unsafe control action is carried out, leading to a loss;
- a correct control action is potentially carried out at the wrong time (too early/late);
- a correct control action is potentially stopped too soon.

Any inadequate controls at each hierarchical level are identified using the interactions established when the safety control structure was constructed (Table 6.6).

Each of these inadequate controls is used to generate constraints and safety requirements to incorporate into each element of the system. Potential inadequate and defective control actions can form the basis for an inventory of the defects and hazards that may cause the system to migrate towards an accident. This inventory can be used to define the constraints that each hierarchical level must satisfy in order to maintain an acceptable level of safety (Table 6.7). These inadequate control actions and constraints are described as *potential* because they are assumed to exist and are only defined in the context of a safety assessment.

Step 5—Risk Categorisation (Immediate, Long Term, Ordinal) and Risk Management (Defective Control Loops)

In this step, the risks that have been identified are initially categorised to determine the impact of inadequate control measures on the behaviour of the system. Risk management involves identifying those processes that can lead to the violation of one or more safety constraints.

Table 6.6 Inadequate control actions for the controller: the operating company

Novosol® operator (SEDISOL or SIFA) **(potential) inadequate control actions**
• The operator does not provide operating requirements to the Novosol® developer to ensure safe operations • The operator does not provide development requirements to the Novosol® developer • The operator does not provide inspection reports to regulatory authorities
• The operator provides incorrect operating requirements to the Novosol® developer • The operator provides incorrect development requirements to the Novosol developer • The operator provides incorrect inspection reports to regulatory authorities
• The operator delays providing key operating requirements to the Novosol® developer • The operator delays providing provide development requirements to the Novosol® developer • The operator delays providing inspection reports to regulatory authorities
• The operator does not provide all operating requirements to the Novosol® developer • The operator does not provide all development requirements to the Novosol® developer • The operator does not provide all inspection reports to regulatory authorities

Table 6.7 (Potential) constraints for the controller: the operating company

Novosol® operator (SEDISOL or SIFA) (potential) constraints
• The operator must provide operating requirements to the Novosol® developer to ensure safe operations • The operator must provide development requirements to the Novosol® developer • The operator must provide inspection reports to regulatory authorities
• The operator must provide correct operating requirements to the Novosol® developer • The operator must provide correct development requirements to the Novosol® developer • The operator must provide correct inspection reports to regulatory authorities
• The operator must provide timely operating requirements to the Novosol® developer • The operator must provide timely development requirements to the Novosol® developer • The operator must provide timely inspection reports to regulatory authorities
• The operator must provide all operating requirements to the Novosol® developer • The operator must provide all development requirements to the Novosol® developer • The operator must provide all inspection reports to regulatory authorities

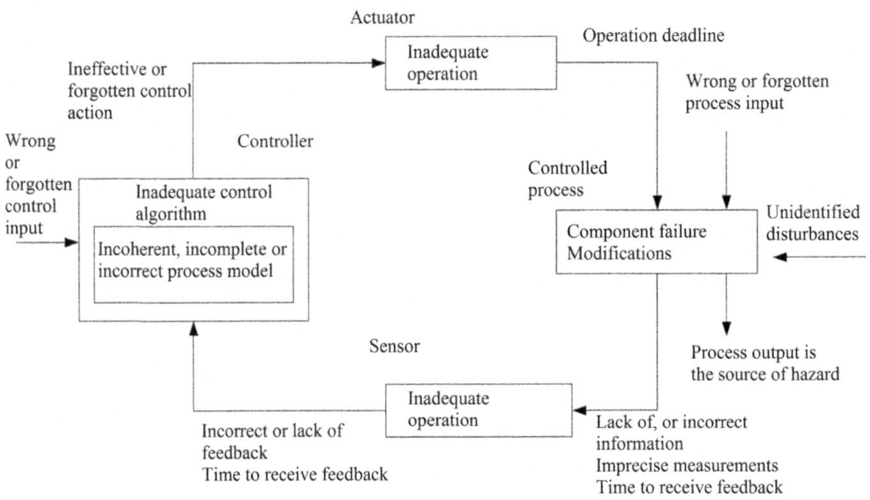

Fig. 6.13 Inadequate control loop. (Leveson 2012)

This step allows control defects to be prioritised. It aims to optimise the safety level of the system by first addressing the immediate risks that could potentially migrate the system towards an accident in a very short time; then the long-term risks that could lead to an accident after a considerable period of time; and finally, so-called *ordinal* risks that are addressed by on-going risk management throughout the system's life cycle. The challenge in this step is to decide which constraints should be given the highest priority.

Once the safety constraints have been identified, it is then necessary to establish where in the control loop they may be violated (Figs. 6.12 and 6.13). The risk management phase of the development of a Novosol® facility begins with the analysis of

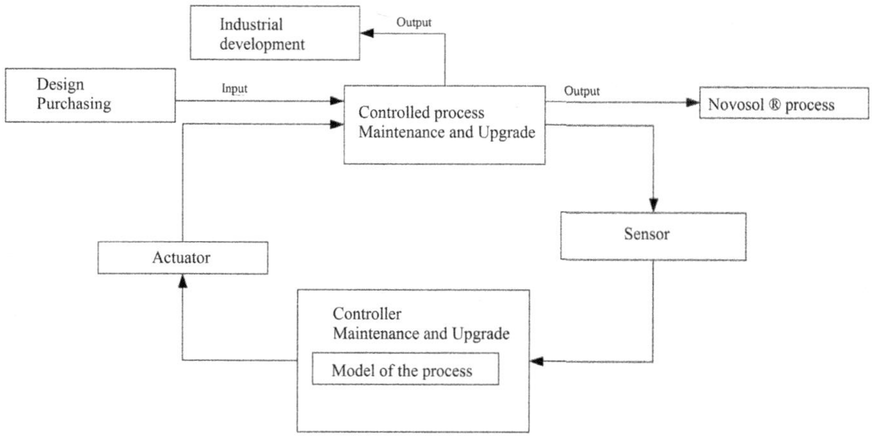

Fig. 6.14 *Maintenance and Upgrade* control loop

existing control loops. There can be inadequate controls at each level of the loop and in each interaction between levels of the loop. The aim is to identify any inadequate controls that can migrate the system to an accidental state, for each of the hierarchical levels. Each of these controls in the loop can give rise to the formulation of an inadequate control at another level, causing the system to migrate towards an accident state.

Any potential actions within the control loop that can lead to an inadequate control need to be identified. Each loop in the control structure, in this case, is part of the Novosol® system and has to be analysed from a systemic perspective (Fig. 6.15) to determine the source of potentially inadequate controls. The control loop for the level, *Maintenance and Upgrade*, for example, can be described in a simplified way (Fig. 6.14), i.e., without establishing potential inadequate controls. This simplified description highlights all of the elements that are involved in the control of the process that interacts with the levels *Industrial Development* and *Design–Purchasing*.

The *Maintenance and Upgrade* control loop, however, may contain incorrect information that could cause an inadequate control at the levels of *Industrial Development* and *Design–Purchasing*. The integration of lower level control loop within the control structure (Fig. 6.15) illustrates how one hierarchical level can interact with the rest of the control structure.

This chapter has shown the contribution of STPA methodology in order to assess risks in the different activities of a sediment decontamination process. Following the accident model proposed by Nancy Leveson, the next chapter presents the application of STAMP methodology to investigate an accident occurred in the field of oil and gas industry.

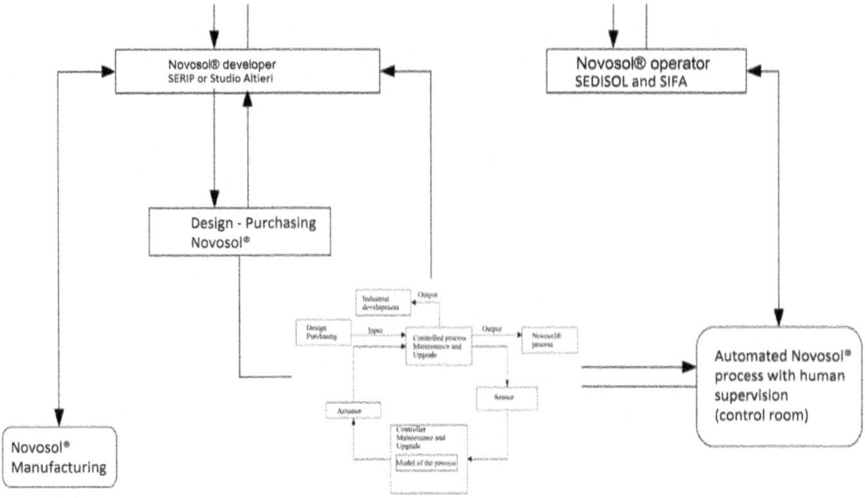

Fig. 6.15 Description of a control loop in the control structure of the Novosol® system

References

Breugelmans D (2007) Novosol®: the story of a pluridisciplinar step by step approach, environmental research and development. Solvay S.A, Belgium

Depelsenaire G (2006) Novosol®: procédé de stabilisation pour résidus minéraux contaminés par des métaux lourds et des composés organiques, Solvay S.A

Ericson CA II (2005) Hazard analysis techniques for system safety. Wiley, New York

Hardy K, Guarnieri F (2011a) Analyzing contaminated sediments through a systemic model of accident. In: Proceedings 41st ESReDA seminar – advances in reliability-based maintenance policies – 41st ESReDA Seminar – advances in reliability-based maintenance policies, France

Hardy K, Guarnieri F (2011b) Applying a systemic model of accident within a system for treatment of contaminated sediments. In: Advances in safety, reliability and risk management: ESREL 2011 – ESREL 2011, France

Leveson NG, Daouk M, Dulac N, Marais K (2004) A systems- theoretic approach to safety engineering: a case study mit esd external symposium. Cambridge, MA

Leveson N, Books24x7 (2012) *Engineering a safer world: systems thinking applied to safety*. The MIT Press, Cambridge, MA

Novosol® (2010) www.novosol.be. Solvay S.A

US-EPA (2005) Contaminated sediment remediation guidance hazardous waste site

von Bertalanffy L (1968) General system theory. Foundations, development, applications. Georges Braziller, New York

Chapter 7
Contribution of the Stamp Model to Accident Analysis: Offloading Operations on a Floating Production Storage and Offloading (FPSO)

Dahlia Oueidat, Thibaut Eude, and Franck Guarnieri

The *System Theoretic Accident Model and Processes* (STAMP) is an approach developed at MIT (Leveson 2011) to analyse accident. STAMP draws upon theories that were developed in the mid-twentieth century in the fields of systems engineering and cybernetics. The approach provides the steps to follow in order to evaluate industrial system and to identify or design at operational level the work tasks performed by its controlling systems (automated, or semi-automated with a human supervisor). The accident analyses approach proposed,in this article, differs from the paradigm normally used in the oil and gas installations in that it provides a comprehensive review of the industrial system. The analyses aims to inspect technical, human and organizational factors involved in an accident. In this chapter, the STAMP model is used to analyse the accident that happened on a *Floating Production Storage and Offloading* (FPSO) unit in the Gulf of Guinea.

FPSO platforms have been used in the offshore oil and gas industry since the 1970s. The FPSO units are usually deployed for exploration and production operations in deep waters. An offshore production platform consists of an infrastructure that is built on the seabed (the *Subsea Production System*) i.e. the subsea oil wells for oil extraction, an FPSO for oil production and storage and a buoy for oil offloading operations between FPSO and oil tanker. The buoy is used, in some oil company, that does not conduct tandem offloading,it is necessary to avoid collision between FPSO and oil tankers during the transfer operations. In fact, the oil tankers connect with the buoy for the loading operation this commercial transaction is subject to customs duty. These infrastructures allow to extract; to process; and to store the oil in the FPSO cargo tanks; the oil stored can be offloaded onto the tanker. The FPSO storage tank system is complex because of the high volumes of liquid involved, procedures

D. Oueidat · T. Eude · F. Guarnieri (✉)
MINES ParisTech/PSL Research University, CRC, Sophia Antipolis Cedex, France
e-mail: dahlia.oueidat@mines-paristech.fr; thibaut.eude@mines-paristech.fr;
franck.guarnieri@mines-paristech.fr

F. Guarnieri, E. Garbolino (eds.), *Safety Dynamics*, Advanced Sciences and Technologies for Security Applications,
https://doi.org/10.1007/978-3-319-96259-7_7

have to be followed to ensure the integrity, buoyancy and stability of the floating unit. All operations (loading, offloading, removal, and internal transfers) must be carried out in accordance with well-defined criteria; any anomalies could lead to unevenly distributed loads, which can damage the FPSO hull. The operators on board are responsible for planning operations and ensuring the smooth running of activities.

The research presented here offers several insights for risk analysis and accident prevention. Several approaches were identified and explored before the *System Theoretic Accident Model and Processes* (STAMP) method (Leveson 2011) was selected. The objective is to systemize the rules, directives and safety measures. The safety constraints must be then implemented according to the hierarchical structure of the organization. This structure is responsible to control the workflow and the enforcement of safety guidelines. These industries are often inspected by governemental authorities, regulatory bodies codify and enforce rules and regulations to ensure safety of the infrustructure, operators, population and the environment. The standards to follow can be prescribed by international convention or can be specific to the country and the type of the industrial activity. The hierarchical structure, to be considered in the analyses, concerns as follow the technical infrastructure, the workers, the organization and the comany administration, the regulatory authorities, the government agencies and the international community. STAMP address the structure responsible for the safety enforcement, supervision and control of the operation in the industry. This organizational structure must controlling the safety of the system functions and comply with the recommended (Oueidat et al. 2015).

7.1 Objectives

This chapter presents the results of the analysis, using STAMP, conducted to understand the accident that happened on an FPSO in the Gulf of Guinea. An uncontrolled leack occured during the commercial transfer of hydrocarbon from the FPSO to an oil tanker. The aim of the analysis was to establish a dynamic organization structure accountable to control and manage the safety of the FPSO and the operations on board. An accident report was drawn up by a group of experts appointed by the company. The method used to identify the causes the accident is considered limited by the scientific comunity because it is based on linear causality reasoning. In the description of the events as they unfold, the experts were unable to precise accurately the principal cause of the accident (the primary event that triggered the accident). STAMP is used to demonstrate that there are factors other than the ones mentionned in the accident report of the experts appointed by the company.

The complete socio-technical system was explored using the framework of the *Causal Analysis based on STAMP* (CAST) accident model. CAST was used to examine the work tasks, process control mechanisms and safety procedures of the controlling systems (technical, operator, managers, organization, regulatory bodies). The objectives of the CAST model is to provide a systematic frame to explain the causes of the accident, it allows to understand managerial flaws, technical failures and downgraded controlling mechanisms. The approach can be then used to re-engineer process control mechanisms, and to design a control structure in charge to enforce the safety constraints that during operational process.

7.2 Methodology

Traditional accident models usually explain the cause of accidents as a chain-of-events. The STAMP method, however, focuses on the notion of *constraints* rather than *events* (Hardy and Guarnieri 2012), and considers an accident to be the result of design flaws and lack of control of the operational process by controlling systems. The accident causes are analysed by exploring the control flaws during operational process. These control flaws usually arise and affect all the development and operations phases. These control problems can be organized, systemized and used during the accident analysis. The company can use these CAST frames of classified control problems to draw a policy and safety plan for accident prevention. This method has been widely used in the analysis of several major industrial accidents (Stoop and Benner 2015; Kwon 2015; Underwood and Waterson 2014; Leveson 2016; Dong 2012; Yang and Tian 2015; Underwood et al. 2016; Hickey 2012; Kim et al. 2016).

A STAMP analysis of accident requires crucial information to fully understand the flaws that result from loss of control. The accident investigation of the Gulf of Guinea FPSO starts with an analysis of the safety instructions that are violated at the operational process. The CAST model, which consists of nine main steps, is then used in the accident analysis. The process can be used by accident field investigation experts (Table 7.1).

These steps are applied to the Gulf of Guinea accident below.

7.2.1 Step 1: Identify System(s) and Hazard(s) Involved in the Loss

The study focuses on the system that ensures the transfer of the crude oil between the FPSO and oil tankers. This system includes two 18.5 inches diameter, flexible transfer lines for the transfer of oil from the FPSO to the offloading buoy anchored 1942 m to the east of the FPSO. The buoy is designed for offloading operation onto Very Large Crude Carrier (with deadweight of 330,000 t) and to avoid collision between vessels.

The FPSO/crude oil transfer system includes the following elements, as shown in Fig. 7.1:

- The FPSO: this can process approximately 250,000 barrels per day (about 40,000 m^3/day), with a storage capacity of about two million barrels of oil.
- The offloading network (for loading operations): this consists of flexible sub-sea lines connected to the buoy, the buoy itself, and the floating offloading line running to the oil tanker. Together these elements comprise the export line (circuit).

Table 7.1 Steps of the applied methodology

Steps	Description
1.	Identify the system(s) and hazard(s) involved in the loss
2.	Identify the system safety constraints and system requirements associated with that hazard.
3.	Document the safety control structure in place to control the hazard and enforce the safety constraints. This structure includes the roles and responsibilities of each component in the structure as well as the controls provided or created to execute their responsibilities and the relevant feedback provided to them to help them do this. This structure may be completed in parallel with the later steps
4.	Determine the proximate events leading to the loss
5.	Analyse the loss at the physical system level. Identify the contribution of each of the following to the events: physical and operational controls, physical failures, dysfunctional interactions, communication and coordination flaws, and unhandled disturbances. Determine why the physical controls in place were ineffective in preventing the hazard
6.	Moving up the levels of the safety control structure, determine how and why each successive higher level allowed or contributed to the inadequate control at the current level. For each system safety constraint, either the responsibility for enforcing it was never assigned to a component in the safety control structure or a component or components did not exercise adequate control to ensure their assigned responsibilities (safety constraints) were enforced in the components below them. Any human decisions or flawed control actions need to be understood in terms of (at least): the information available to the decision maker as well as any required information that was not available, the behavior-shaping mechanisms (the context and influences on the decision making process), the value structures underlying the decision, and any flaws in the process models of those making the decisions and why those flaws existed.
7.	Examine overall coordination and communication contributors to the loss
8.	Determine the dynamics and changes in the system and the safety control structure relating to the loss and any weakening of the safety control structure over time
9.	Generate recommendations. In general, the description of the role of each component in the control structure will include the following: Safety Requirements and Constraints Controls Context Roles and responsibilities Environmental and behavior-shaping factors Dysfunctional interactions, failures, and flawed decisions leading to erroneous control actions Reasons for the flawed control actions and dysfunctional interactions Control algorithm flaws Incorrect process or interface models. Inadequate coordination or communication among multiple controllers Reference channel flaws Feedback flaws

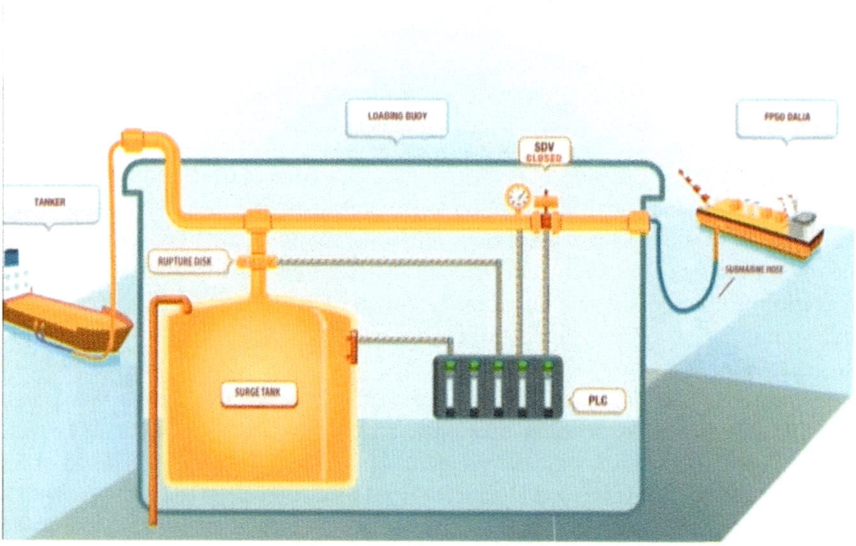

Fig. 7.1 Offloading operation from the FPSO to the oil tanker via a offloading buoy

- The offloading buoy: the buoy is fitted with a 100 m^3 Surge tank, which is used in an emergency to collect oil of transfer lines if there is a problem during offloading operations. This tank is designed to hold any hydrocarbons that may be in transit during the time (mechanical response delay) the Shut-Down Valve (SDV) closes. The Shut-Down Valve (SDV) is an emergency stop valve that is used to isolate the export line with the tanker under certain conditions to avoid an accidental pollution.
- If the pressure in the circuit reaches 15 bar, a rupture disk (a safety device, a kind of blank flange, that isolates the *Surge Tank* from the circuit and has a predetermined mechanical strength) breaks and the *Surge Tank* is filled to avoid accidental pollution.
- A Process Logical Controller (PLC) is installed on the buoy for the purposes of closing the Shut-Down Valve (SDV).SDV can be operated also from the control room. The SDV is closed if the pressure in the export line reaches 10 bars, or the volume of oil in the *Surge tank* is high (as evidenced by the broken rupture disk). The PLC operates autonomously; any local anomalies that are detected on the tank lead to the closure of the SDV without going through the FPSO control system.
- A fibre optic cable for digital data transmission of the buoy parameters to the FPSO control room. Operators in the control room can stop the offloading operations if there is a problem on the buoy.
- The oil tanker: once moored and connected to the buoy, it receives hydrocarbons from the bottom of the FPSO tanks (the ones selected for offloading by the loading master). This stage is known as *de-bottoming* and is primarily used to separate, if necessary, water (notably from production wells and the oil decantation),

but also to check the layout of the circuit and its tightness. This important step is often done at low flow rates. Offloading is proceed at a nominal rate until the requested amount has been loaded into the tanker. At the final phase (completion or topping up), the flow rate is reduced again to avoid any overflow.

This accident is due to the damages to the fibre optic cable that transfer the parameters of the offloading buoy which have not been transmitted to the FPSO control room. A technical intervention was planned to install a temporary radio-transmission system, as a replacement, this intervention failed because the radio link didn't operate effectively and transmit the required data. During the technical intervention, the technicians temporarily cut the electricity supply to install the radio-transmission system, the loss of electric power caused the SDV to close. This closure was not reported and went undetected because it was outside the remit of telecom technicians. The day of the accident, the oil transfer circuits was being prepared for the tanker loading operation, an operator noted that the SDV was closed. The SDV position was interpreted as an unexpected closing. Thus in order to prevent the SDV unexpected closing, it was "electronically" forced open with a manual input in the PLC (an override). This operation is dangerous as it inhibits the safety control functions of the PLC and the control room operator.

7.2.2 Step 2: Safety Constraints in the System and Required Configurations

During offloading, the tanker's pilot (or his assistant), the FPSO Operations Manager (by delegation the FPSO loading master) and the tanker's Loading Master coordinate operation phases by radio communications means. The tanker's Loading Master is the person in charge of the preparation and monitoring of loading operations, he is the guarantor from the client side of the oil load quality/quantity; there is also a Loading Master on the FPSO side who fulfils the same functions.

Instructions are concurred between the FPSO and the tanker's pilot and/or the Berthing Master. Any changes to the configuration of the valves on the tanker's offloading circuit must be notified to the tanker's pilot, who informs the FPSO's Operations Manager in order to avoid any risk of over pressurization in the circuit. The tanker's Loading Master acts as a counsel during the offloading operations, as the person responsible (according to the law and regulations) is still the master of the ship and by delegation, the chief officer for the cargo operations and the safety on board. An emergency stop can be triggered if there is a problem with the mooring or the connection, an uncontrolled leak of hydrocarbons, or a major accident. In this case, the tanker's Loading Master must warn the pilot before he closed the valve at the manifold. The pilot, in turn, informed the FPSO.

The safety constraints applied to offloading operations are:

- The transfer of hydrocarbon is continuously under positive control.
- Precautionary measures must be in place to protect the infrastructure and its equipment.

- Precautionary measures must be in place to protect the environment.
- Precautionary measures must be in place to minimize human casualties and material damage, if there is an unexpected fire or explosion.

7.2.3 Step 3: The Hierarchical Safety Control Structure

The CAST method requires to configure the safety control structure in charge of operational process. Figure 7.2 illustrates the control structure, the entities in this structure are in charge of safety enforcement during system phases. This structure includes international and official entities, states, certification agencies, and international organization for standardization that promulgate the regulations and requirements for safety enforcement in the maritime field. The operators are responsible for ensuring the safety of the operations and the environment. This structure makes it possible to visualize the interactions between various system components.

Figure 7.2 illustrates the structure that ensures the enforcement of safety constraints and feedback on operational process. During accident investigation, control flaws are inspected for each entity of this structure. At the theoretical level, the safety control directives that are presented in the figure must be compliant with the requirements of:

- The International Maritime Organization (IMO). This United Nations body is responsible for the administration of the sea and maritime navigation, and draws up conventions. The most important are:
  - The 1974 International Convention for the Safety of Life at Sea, as amended.
  - The 1973 International Convention for the Prevention of Pollution from Ships, as amended by the Protocols of 1978 and 1997.
  - The 1978 International Convention on Standards of Training, Certification and Watch keeping for Seafarers, as amended, including the amendments of 1995 and the Manila Amendments of 2010.
- Governments, who adopt international regulations that aim to ensure maritime safety. Official certification bodies inspect facilities and deliver certificates of compliance with standards and international regulations.

Strictly speaking, there is no general requirement for a permanently moored F(P)SO installation to be flagged or to comply with international maritime law, as it is not a vessel undertaking international voyages. However, in cases where an F(P)SO will proceed to site under its own power or where it may be taken off station a flag may be desirable or necessary. In addition, some Coastal states require that F(P)SOs operating in their waters be Flagged, either under their own jurisdiction, or under another Flag State and that they are issued with international statutory certification as if they were on international voyages. Flagging a vessel places the responsibility on the operator of ensuring compliance with international marine standards as required by the administration of the Flag State. The regulations are generally those imposed by the International Maritime Organization (IMO), an agency of the

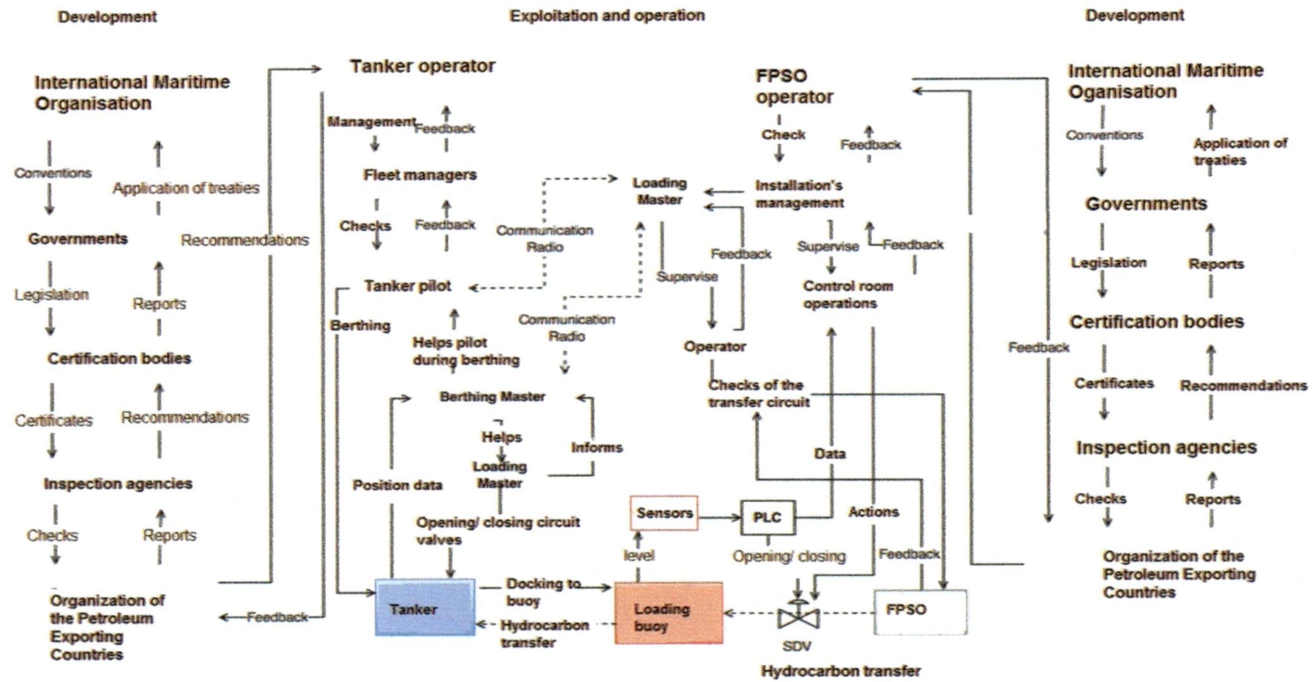

Fig. 7.2 The control structure. In Fig. 7.2, the arrows represent the control and feedback between control systems, the dotted arrows represent the communication or transfer channels

United Nations. Flag administrations apply international statutory requirements directly or through national agencies, which may ask for additional requirements (OGP Safety Committee 2006). For compliance with Flag state requirements, F(P) SOs may require the issue of at least the following certification:

- Safety Construction Certificate (SOLAS 1974)
- International Oil Pollution Certificate (MARPOL 73/78)
- International Loadline Certificate (1966)
- International Tonnage Certificate (1969)
- MODU Safety Certificate (MODU Code 1989)
- Shipboard Safety Management Certificate (ISM code)
- Certificates for Masters, Officers or Ratings

These certificates would be issued by the Flag State or by Class on behalf of the Flag State. The Flag state would also expect the following certificates to be issued by the Class Society supported by appropriate yard and manufacturers' documentation:

- Certificate of Class (hull, machinery, etc.)
- Certificates for lifting equipment/cranes.
- In Europe, compliance with applicable directives (F(P)SOs will generally be considered to be offshore installations and not ships)

Various control systems serve to guarantee the safety of operations at the technical level. Several controllers may impose the same guidelines or safety constraints.

At the technical level (Fig. 7.3), the operational processes are the FPSO's cargo tanks, the SDV, and the offloading buoy. The tanker's Loading Master is responsible for preparing the loading sequence, while the Utility Operator (the Loading Master's point of contact in the control room) is responsible for implementing the plan under the Loading Master's supervision.

7.2.4 *Step 4: The Facts of the Accident*

The accident investigation internal report describes how events unfolded: a large oil spill into the sea occurred at the FPSO's buoy. The overall offloading circuit between the FPSO and the tanker is protected from overpressure by the rupture disk that is installed on the offloading buoy (Fig. 7.1). The rupture disk is calibrated with a bursting pressure of 15 bars and varies between a minimum of 14.3 bars and a maximum of 15.8 bars.

Many dangerous events began to unfold five months before the accident. It appears that the system was working in a downgraded mode because the communication link between the FPSO and the buoy had broken down due to damage to the fiber optic cable that ensures the connectivity. On the day of the accident, the buoy's safety system was operating in local mode, but it was impossible to operate the SDV. There was, however, no way to automatically stop the offloading operation

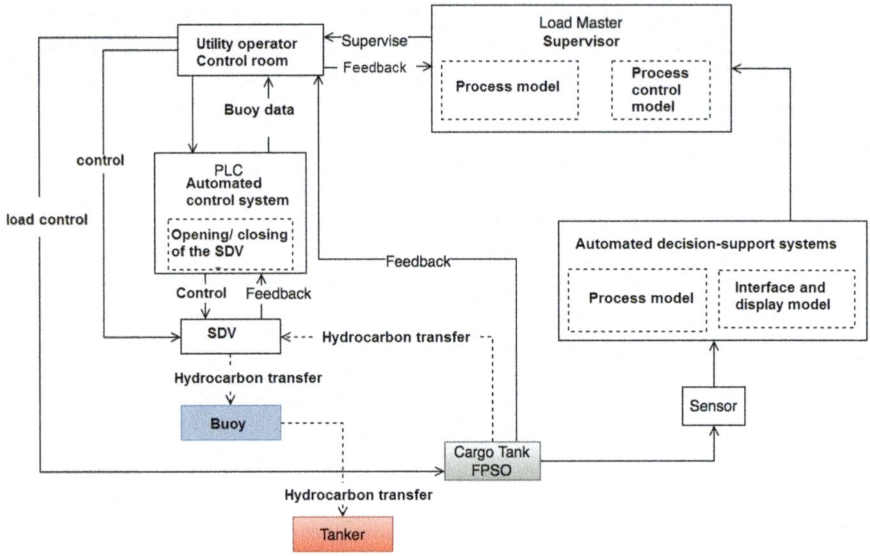

Fig. 7.3 Model of the control structure for tanker offloading. In Fig. 7.3, the boxes represent the systems, the dotted boxes represent the functions that can be issued by a system and its interfaces display. The dotted arrow shows an example of the flow that can be transferred between systems

from the FPSO control room if the buoy's oil volume became too high in the surge tank. Although overpressure on the buoy can lead to the closure of the SDV, offloading could only be halted if a peak of pressure occurred at the outlet side of the FPSO cargo pumps (i.e., after the SDV was already closed).

The downgraded mode of operation was only identified ten days after the fiber optic cable damage. A technical intervention was carried out, which consisted of connecting a temporary radio-transmission system to the buoy. A short electrical power cut caused the SDV to close. This closure was not noticed by the technician installing the radio-transmission system or the control room.

A few months earlier, when an offloading operation was scheduled, an operator noticed that the circuit and the SDV were closed. In order to prevent it being unexpectedly closed, they decided to force the valve open with a specific output electronic signal; in other words, to keep it physically open. An intervention request to fix the problem was prepared and approved, but the initial, downgraded situation was not addressed. Fifteen days later, a test of the new radio-transmission link between the buoy and the FPSO was carried out: it proved unsuccessful because communication was random and data appeared frozen on the screen. The SDV was once again forced into an open position.

On the day of the accident, offloading operations began at the end of the afternoon. In the evening, a peak pressure of at least 14.7 bar was noted at the FPSO's discharge pump (this pressure was probably due to the closing of cargo valves on the tanker as offloading was coming to an end). Following this increase in pressure, the rupture disc broke, the Surge Tank on the buoy filled, and then overflowed. The PLC, which was designed to activate the closure of the SDV had been disabled. Shortly afterwards, the tanker's crew noted a fall in the flow rate and this anomaly was flagged up and recorded. Unfortunately, no further action was taken and the details of the problem were not passed to the FPSO. A quarter of an hour after the disk ruptured, the drop in the offloading rate was noted by the tanker's Loading Master, who also flagged up and recorded the anomaly. However, he did not alert the FPSO's Loading Master.

The FPSO control room did not detect anything unusual: data from the buoy were frozen, but recordings of the transfer parameters initially showed a drop in the requested flow rate (consistent with the end of the loading operation), followed by a peak in pressure, and finally a flow rate that returned to its initial value, accompanied by a subsequent fall in pressure. But these signals were ignored and the event went undetected. Finally, during the night, the smell of oil alerted a member of the tanker's crew, and pumping was halted. An unknown quantity of crude oil was dumped into the sea, more than 80 km from the coast. A crisis unit was set up and operations to clear up the spillage began. They would last almost 3 weeks.

7.2.5 *Step 5: Analysis of the Weaknesses of the Technical Installation*

The causes of the accident are analysed by first collecting data to identify the risks that are inherent to the loss of control. To do this, the technical failures that led to the accident are analysed. It is important to identify how the following control mechanisms contributed to the progress of the accident: control of the technical installation, repairs, failures, and communication, and weaknesses in unhandled disturbances. It is also important to document and explain why these control mechanisms are inappropriate and played an inherent part in the oil spillage. By following this approach, a risk prevention plan can be developed and then implemented.

7.2.5.1 Inappropriate Control and Feedback to Overcome the Problem of the Severed Fibre Optic Cable

The reasons for the failure of the fibre optic cable are not analysed in the investigation report. It is important to understand this failure to prevent similar incidents from happening in the future. An external company intervened to connect a temporary remote radio transmission system (3 weeks after the fibre optic cable was severed) to allow the transfer of data from the buoy to the control room. This intervention did not fix the problem, and data transmission was not restored. Ten days later, the external company carried out a second intervention, which also failed to solve the telecommunication problem.

7.2.5.2 Inappropriate Control and Feedback to Force the SDV Open

During a visit to the buoy in preparation for the loading operation, the SDV was found to be closed. An informal analysis of the situation led to the decision to force it open in case it unexpectedly closed during the loading leading to hydraulic shock. This intervention meant that the SDV could no longer be opened or closed by the local PLC safety controller and the signals that powered the PLC (*Surge Tank* level, *Surge Tank* pressure, rupture disk status). It also meant the valve could not be opened or closed by sending instructions from the control room.

7.2.5.3 Inappropriate Control and Feedback Following the Breaking of the Rupture Disk

The breaking of the rupture disk led to the filling and overflow (via the overflow pipe) of the Surge Tank on the buoy. No alerts were received in the FPSO control room, and there was no automatic closure of the SDV or the safety valves in the FPSO cargo circuit. The flow and the leak were stopped at 9.54 pm at the tanker pilot's request, after a crew member noticed a smell of crude oil, which was suspected to be linked to pollution.

7.2.6 Step 6: Component Analysis of the Hierarchical Structure

After the analysis and identification of the factors that contributed to the loss of control on the physical installation, the next step consists of successively examining the hierarchical levels. This involves understanding the mechanisms that led to the loss of control in the operational process. For each component, the conduits that foster the propagation of an inadequate control have to be identified. Each system safety instruction is checked to see if it had been assigned to a competent controller, and it was properly executed at the operational process. To do this, all of the decisions and failures in the actions that were taken have to be examined using the information that was available to the decision-makers, (note should be taken for necessary information that was unavailable). The analyses must include actions (the context and influences on the decision-making process) and the reasoning structures that underlie the decisions that were made, and explain why these flaws occurred.

The CAST accident analysis approach is applied in a bottom-up manner, and the behaviour of frontline operators is studied first. Then we must go back up the control structure to list the weacknesses that contributed to the accident (Fig. 7.4) of each controlling system at the operational process. In what follows, we simply present the analysis of the control room operator and the front-line operator.

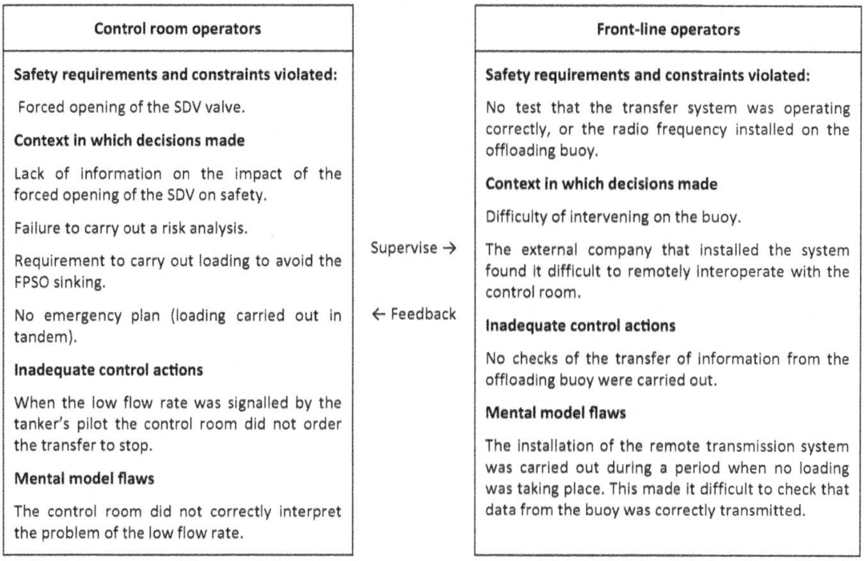

Fig. 7.4 Analysis of operator behaviour

7.2.7 *Step 7: Evaluation of Coordination, Communication and Control of the Components in the Hierarchical Structure*

The analysis developed in step 6 consists of evaluating each component in the hierarchical control structure. In step 7, the coordination and communication between controllers and decision-makers is analysed. The approach identifies potential conflicts and coordination problems, and examines side effects of decisions or actions or conflicting control actions. The analysis highlights many coordination and communication weaknesses.

7.2.7.1 Inappropriate Coordination and Communication Led to Weaknesses in the Hierarchical Management of Maintenance

The procedure to connect the temporary remote radio transmission system shows flaw in verifying and validating the conducted intervention by the communication technicians. It shows uneffective coordination and communication among controllers and decision-makers. Indeed, signal verification tests were not correctly prepared. Cutting the power to make the connection caused the closure of the SDV. This closure was not registered or notified because it was out of scope of the telecommunication technicians.

7.2.7.2 Inappropriate Coordination and Communication

Two hypotheses, related to the accident, arrise for the valve configuration on the tanker's loading circuit:

- The tanker operator does not appear to notify the FPSO operations manager of the configuration of the valve in the tanker's loading circuit at the end of the operation. Recordings show a peak pressure of 14.7 bar on the FPSO export pumps. The control weakness can be interpreted as a lack of coordination during interventions at different levels of the structure (closure of the valves on the tanker at the end of loading operations, a lack of exchange and transmission of control parameters).
- The tanker operator informed the FPSO operations manager of the configuration of the valve on the loading circuit, but the operator in the FPSO control room did not stop the transfer when a peak pressure of 14.7 bar was measured on the FPSO export pumps.

7.2.7.3 Inappropriate Coordination and Communication During Loading

The loading operation was carried out in degraded mode, and the structure responsible for monitoring the safety of the system must be particularly vigilant. The Loading Master on the loading tanker noted the abnormally low transfer rate, however, this warning was ignored by the FPSO team, who did not stop the transfer.

7.2.8 Step 8: Modelling the Accident with the System Dynamic

The system dynamics approach created in 1950s by Forrester, it provides a framework for dealing with dynamic complexity, where cause and effect are not obviously related. Like the STAMP models, it is grounded in the theory of non-linear dynamics and feedback control, but also draws upon cognitive and social psychology, organization theory, economics, and other social sciences. Figure 7.5 shows a simple systems dynamics model of the FPSO accident, this model is only a hint of what a complete model might contain. The figure represents an example of the relationship between state variables that lead to an accident.

Thus each component of the control structure is in charge to maintain the safety of operational process. Communication and coordination play an important role in the enforcement of safety constraints regulated by control systems. This step helps to explain how the decision-making context affects safety control functions over time. The Figure shows a simplified model of the accident, in which the integrity of the technical installation is not properly managed.

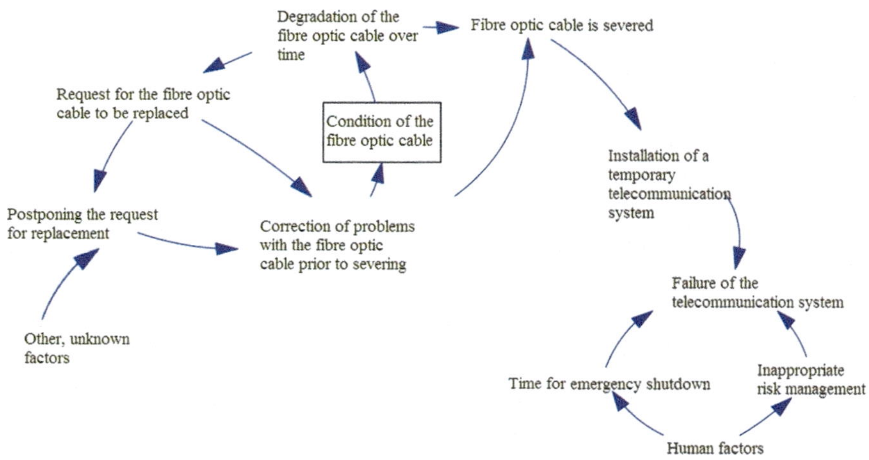

Fig. 7.5 Simplified model of the accident

The severing of the fibre optic cable and the failure of the remote transmission system led to a delay the emergency shutdown.

7.2.9 Step 9: Recommendations

In CAST, the aim of the accident analysis is not to accuse or blame the weakest member of the control hierarchy. Instead, it is to draw lessons that can be applied in order to improve operations and safety guidelines of operational process. A comprehensive CAST analysis promotes a set of recommendations. For the FPSO accident, they are classified into four categories:

7.2.9.1 The Technical Infrastructure

- The SDV should be placed under local control by the buoy's PLC. Operators must supervise operations, and loading parameters, in order to detect any indicators of a loss of control at the earliest possible opportunity.
- The operator must strengthen the valve control system.
- Other alerts must be added to supplement the computer system.
- Strengthen direct physical perceptions of operators. The control room operators of the FPSO were unaware that anything was wrong as control parameters were only provided from the PLC, and following the breakdown in communication the control system was defective.
- Provide a reliable backup communication system (to avoid degraded functioning and a loss of control).

7.2.9.2 Business Management

In the case of the FPSO, there is the oil company and the shipping company to consider. Each must ensure that their respective facilities (the FPSO and oil tanker) comply with the regulations, norms and standards that apply to their activities. During loading operations, which lie at the interface between the two actors, it is necessary to maintain safety process control (e.g., real-time coordination between the FPSO operators and the tanker's crew, supervision of the automated functions of the PLC).

For this it is necessary to establish an internal safety policy, which clearly defines:

- The roles, powers and responsibilities of the different actors in the safety structure.
- The criteria used to evaluate decisions, the design, and implementation of safety controls.
- The requirements for the systematic application of safety constraints.
- The organization responsible for the process control, which must ensure:
  - The application of the strategy;
  - Senior managers are made aware of safety decisions;
  - Risk analyses and audits are properly recorded;
  - Safety constraints are defined as a function of the activity, and changes are registered and documented;
  - Make available to the experts a reference document for accident investigations based on CAST;
  - The establishment of a systemized information system specific to operational process safety control;
  - The establishment of a coordination structure for information sharing and cooperation between the different actors.

7.2.9.3 Management and Production

There should be a proper safety strategy specific to the infrastructure (the *Physical Plant*) in addition to one for the company:

- To conduct risk analyses;
- To investigate the reasons and conditions for incidents;
- To evaluate risk indicators;
- To systematically collect data;
- To validate the training programs of operators.
- To compel criteria to be adopted for safety control:
  - Mandate the systematic application of safety constraints;
  - Enhance communication and transparency;
  - Ensure that safety measures are validated by the head of the infrastructure team;
  - Ensure communication between the various components of the infrastructure.

In this case, loading began in degraded mode (no check was made on the connection of the radio link, there was no reaction to the frozen data, there was a lack of communication between the tanker's operators and those on the FPSO, etc.).

7.2.9.4 Government and Environment

The aim is not to isolate the system from its social environment or to neglect its commitments towards governments and legislation governing safety control.

Using CAST, in this article we propose a frame to describe unsafe decisions and control actions adopted by each componant of the structure. Based on STAMP foundations, inadequate coordination and dysfunctional communication ans interactions are analysed (e.g., defective cognitive models, lack of coordination between controllers, flaws in instruction execution flaws in feedback). By using this approach, it becomes possible to study behaviour shaping mechanisms (Fig. 7.5), to understand the decision-making process that led to the accident. A more in-depth CAST analysis, however, is limited in the sense that: (1) it requires a wide variety of data associated of the overall system that are difficult to obtain; (2) the operator may find it difficult to implement the recommendations resulting from the analysis in a timely manner. Concerning this case study, there are clearly unresolved issues, although here we propose new ideas that open up ways to improve our understanding of accidents in the maritime fields.

The next chapter presents an application of system dynamics thinking in the field of risk management for CO_2 capture, transport and storage projects.

References

Dong A (2012) Application of CAST and STPA to railroad safety in China. Massachusetts Institute of Technology. http://dspace.mit.edu/handle/1721.1/76491

Hardy K, Guarnieri F (2012) Modéliser les accidents et les catastrophes industrielles. http://www.lavoisier.fr/livre/sciences-du-risque/modeliser-les-accidents-et-les-catastrophes-industrielles/hardy/descriptif-9782743014773

Hickey J (2012) A system theoretic safety analysis of U.S. Coast Guard Aviation Mishap Involving CG-6505. Thesis, Massachusetts Institute of Technology

Kim T-e, Nazir S, Øvergård KI (2016) A STAMP-based causal analysis of the Korean Sewol ferry accident. Safety Science 83:93–101. https://doi.org/10.1016/j.ssci.2015.11.014

Kwon Y (2015) System theoretic safety analysis of the Sewol-Ho Ferry Accident in South Korea. Massachusetts Institute of Technology. http://sunnyday.mit.edu/papers/Kwon-Thesis.pdf.

Leveson N (2011) Engineering a safer world: systems thinking applied to safety. https://mitpress.mit.edu/sites/default/files/titles/free_download/9780262016629_Engineering_a_Safer_World.pdf.

Leveson N (2016) Rasmussen's legacy: a paradigm change in engineering for safety. Appl Ergon. https://doi.org/10.1016/j.apergo.2016.01.015

OGP Safety Committee (2006) Guideline for managing marine risks associated with FPSOs. http://www.ogp.org.uk/pubs/377.pdf.

Oueidat D, Guarnieri F, Garbolino E, Rigaud E (2015) Evaluating the safety operations procedures of an LPG storage and distribution plant with STAMP. Procedia Eng 128:83–92

Stoop J, Benner L Jr (2015) What do STAMP-based analysts expect from safety investigations? Procedia Eng 128:93–102. https://doi.org/10.1016/j.proeng.2015.11.508

Underwood P, Waterson P (2014) Systems thinking, the Swiss Cheese Model and accident analysis: a comparative systemic analysis of the Grayrigg Train Derailment using the ATSB, AcciMap and STAMP models. Accid Anal Prev 68(juillet):75–94. https://doi.org/10.1016/j.aap.2013.07.027

Underwood P, Waterson P, Braithwaite G (2016) 'Accident investigation in the wild' – a small-scale, field-based evaluation of the STAMP method for accident analysis. Saf Sci 82(février):129–143. https://doi.org/10.1016/j.ssci.2015.08.014

Yang Q, Tian J (2015) A formal approach to causal analysis based on STAMP (CAST). In http://ieeexplore.ieee.org/stamp/stamp.jsp?tp=&arnumber=7366434

Chapter 8
Systemic Risk Management Approach for CTSC Projects

Jaleh Samadi and Emmanuel Garbolino

Capture, Transport and Storage of CO_2 (CTSC) is a technology contributing to industrial CO_2 emissions and climate change mitigation. CTSC consists of a chain of processes to collect or capture a CO_2 gas stream, transport the CO_2 to a storage location and inject it into that location. The most significant source of CO_2 emissions is the combustion of fossil fuels such as coal, oil and gas in power plants, automobiles and industrial facilities. A number of specialized industrial production processes and product uses such as mineral production (cement, lime, etc.), metal production (iron and steel, aluminum, etc.) and the use of petroleum-based products can also lead to CO_2 emissions (US-EPA 2010).

CTSC is currently a constituent of global energy policy, although there are still lots of uncertainties regarding CTSC contribution and development.

The global target is to halve the current CO_2 emissions by 2050 (GCCSI 2011). International Energy Agency (IEA) estimates that CTSC will reduce 19% of CO_2 emissions by 2050 (IEA 2010).

United Nations Framework Convention on Climate Change (UNFCCC) has emphasized on the importance and urgency of climate change concerns (UNFCCC 2012). However, national policies seem to deal with several uncertainties. Canada's withdrawal from Kyoto protocol just after the last climate change conference in Durban (November 28–December 11, 2011) is an example of uncertain policies.

Perceptions of stakeholders on the effectiveness of CTSC are different. Although most of governments and industries intend to invest on the technology, others such as local communities and NGOs are worried about long term risks and reliability of CO_2 storage. CO_2 leakage is the most significant concern of these groups since it could lead to risks for human beings, animals and plants as well as potable water networks.

J. Samadi · E. Garbolino (✉)
MINES ParisTech/PSL Research University, CRC, Sophia Antipolis Cedex, France
e-mail: jaleh.samadi@mines-paristech.fr; emmanuel.garbolino@mines-paristech.fr

F. Guarnieri, E. Garbolino (eds.), *Safety Dynamics*, Advanced Sciences and Technologies for Security Applications,
https://doi.org/10.1007/978-3-319-96259-7_8

Risk Assessment and Management are essential parts of CTSC development in order to provide answers for the uncertainties and assure the control of well-understood parts of CTSC processes.

Several studies have been carried out on risk assessment of Capture, Transport and Storage technologies. Risks of CO_2 Capture and Transport are assumed to be well understood. Therefore, classical methods have been usually applied for analyzing risks of Capture and Transport subsystems. However, CO_2 storage is known as a "*nonengineered*" part of the process, dealing with various uncertainties (Koornneef et al. 2012). Consequently, most of available risk assessment studies are focused on CO_2 storage technical aspects of risk.

Most of available approaches ignore that CTSC is a complex sociotechnical system for which risks could not be analyzed individually, without taking the whole context into account. Complex system is a *system composed of many parts that interact with and adapt each other. In most cases, the behavior of such systems cannot be adequately understood by only studying their component parts. This is because the behavior of such systems arises through the interactions among those parts* (IRGC 2010). A sociotechnical system is a one consists of a technical part which is in interaction with a social part.

Risks associated to CTSC are not limited to technical risks. Along with technical challenges, CTSC is faced to uncertainties concerning development up to commercial scales. At the present time, 60 large scale integrated projects are identified in the world (GCCSI 2014). The number of projects has been reduced, especially in Europe (74 in 2012 vs. 60 today). A number of projects have been cancelled or delayed for various reasons.

Therefore, a major question about CTSC at the current scale of development is what are the factors explaining the success or failure of CTSC projects in different contexts?

In order to answer this question, a systemic risk management framework is proposed based on the concepts of System Dynamics and STAMP (Systems-Theoretic Accident Model and Processes), developed at Complex Systems Research Laboratory of Massachusetts Institute of Technology.

In this work, systemic modeling is proposed as a decision making support, which provides the grounds of thinking about the components of a potentially successful CTSC project. Each stakeholder is assumed as a "*controller*", who is responsible for maintaining safety constraints. Safety control structures are developed for several case studies to formalize the relations of stakeholders in maintaining safety constraints.

8.1 CTSC & Risk Management

In order to understand why a systemic risk management framework is required for CTSC chain, different aspects of risk should be firstly understood and studied.

Eight major groups of risk are identified in this work for CTSC whole system:

8.1.1 Risks

There are different risks associated to each subsystem. Examples for CO_2 capture include production of vent gas and liquid or solid wastes. Leakage and corrosion are the most significant issues related to CO_2 transport. Risks for human beings, animals and plants above ground, contamination of potable water and CO_2 atmospheric release are examples of CO_2 storage risks.

8.1.2 Risks Related to CTSC Project

Mainly include the risks that affect the project progress, particularly the risks related to the project schedule, cost and performance; and development to commercial scales.

8.1.3 Social (Public Acceptance) Risks

Public acceptance is a risk that could significantly affect CTSC projects development. An example is Barendrecht project, in the Netherlands, which was cancelled due to public disagreement (CCJ 2010). De Coninck et al. believe that *the companies are not worried that CO_2 capture and storage will fail for technical reasons. One of the concerns, however, is potential public resistance to CCS, and some companies indicate that governments should step in to provide neutral information to the lay public* and *it is imperative to find a common language for the characterization and communication of risk both among professionals and between professionals and the public* (De Coninck et al. 2009).

8.1.4 Policy/Strategy Risks

Policy uncertainties are defined as a major risk to CTSC projects development. GCCSI defines four policy landscapes that affect CTSC technology (Fig. 8.1).

Policies are not the same in different countries, and are strictly dependent of the policies regarding Climate Change. Canada's withdrawal from Kyoto protocol after the last climate change conference in Durban, held at the end of 2011, is an example of changing policies.

Nevertheless, the policy making of CTSC is a complex issue, depending on several points.

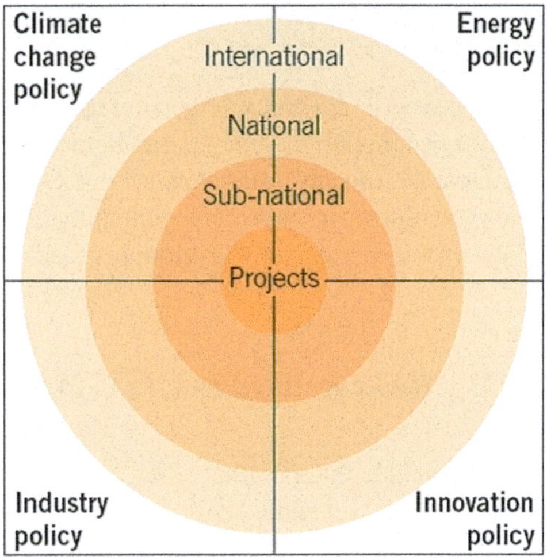

Fig. 8.1 Scope of policy landscapes related to CTSC. (GCCSI 2011)

8.1.5 Health, Safety and Environmental (HSE) Risks

Technical matters, notably impurities, leakage and corrosion may lead to HSE problems.

Carbon dioxide is a colorless, odorless, harmless, non-flammable gas. Impacts on respiratory and cardiovascular systems increase with concentration amplification. There are standards for occupational exposure limits for CO_2.

8.1.6 Regulatory or Legal Risks

According to a survey committed by GCCSI, regulatory issues are a significant challenge for CTSC projects (GCCSI 2011). Several international and regional regulations could cover the requirements of CTSC technology. These regulations need to be transposed into national or domestic laws. De Coninck et al. argue that IPPC Directive (96/61/EC, as amended) is applicable for CO_2 Capture in Europe. The IPPC Directive is the European Commission Directive on industrial emissions. The authors point out that *liquefied CO_2 is already transported in significant quantities by road, ship and pipeline across the EU and is regulated in accordance with dangerous goods laws and regulations*. However, Environmental Impact Assessment Directive (85/337/EEC, as amended) of European Commission could be applied for pipelines and pumping stations. As mentioned by De Coninck et al., EU Directive on CTSC *does not sufficiently deal with all legal uncertainties concerning the capture and transport of CO_2 derived from CCS facilities*. In spite of the availability of EU Directive for CTSC, *under current European law, it is uncertain whether CO_2*

that is captured and then stored would be classified as 'waste'. If so, the European waste laws could be applicable for CO_2 storage. This concern is currently the subject of several research studies. (De Coninck et al. 2009).

8.1.7 Organizational and Human Risks

CTSC is a complex sociotechnical system which includes not only three technical components of Capture, Transport and Storage, but also an organizational structure containing a group of actors. The organizational and human risks are derived from such a complexity.

8.1.8 Financial/Economic Risks

Some CTSC projects have been cancelled due to financial issues. So, financial aspects of project are an important part of risk management studies.

Considering such an overview of risks associated to CTSC activities, a methodology is developed for CTSC risk management. The methodology is detailed in next sections.

8.2 Requirement of a Systemic Approach

CTSC is a complex sociotechnical system which includes a technical system with three components of Capture, Transport and Storage. The social part of CTSC sociotechnical system involves an organizational structure containing a group of actors. The interface between organizational, human and technical aspects could initiate a failure in the system.

Complex systems have some common characteristics including *Emergence, Non-linearity, Inertia, Threshold behavior, and Hysteresis and Path Dependency*. These characteristics lead to difficulties in anticipating and controlling system behavior. (IRGC 2010).

The components of sociotechnical system include human beings (workers, managers and all the stakeholders of internal and external environment), an organizational structure and a technical section (including equipment, methods and tools) (Carayon 2006). These components are in interrelation with the external environment of the system.

With these definitions, we could consider CTSC as a complex sociotechnical system for which novel systemic risk management approaches are required.

OECD (Organisation for Economic Co-operation and Development) confirms that new risk management approaches are required for emerging technologies. New methodologies for analyzing emerging technological risk should be systemic. (Hellström 2009; OECD 2003).

8.3 Available Risk Management Approaches for CTSC Chain

Most of available risk assessment or management methods are focused on one subsystem (Capture, Transport or Storage). Due to the uncertainties concerning the reliability of CO_2 storage, risk analysis studies particularly concentrate on the storage. A great number of studies only analyze the technical risks, and believe that the risks of Capture and Transport could be studied by classic methods.

However, there are examples in the literature that highlight the necessity of an integrative risk management method for CTSC.

Farret et al. underline the importance of developing an integrated approach in risk analysis of CTSC due to interdependency of four steps, i.e. Capture, Transport, Injection to the reservoir and Long-term Storage. (Farret et al. 2009).

Gerstenberger et al. believe that a comprehensive risk assessment method does not yet exist for CTSC and needs to be developed. (Gerstenberger et al. 2009).

The (semi-)integrated studies that have been already carried out on CTSC risk management are as following:

GCCSI applies the Australian and New Zealand Standard for Risk Management (AS/NZS 4360: 2004) to define the likelihood and consequences of a set of extreme risks associated to integrated CTSC projects. Seventeen risks are identified with public, governmental/regulatory/policy, business case and technical nature. (GCCSI 2009).

Det Norske Veritas (DNV) has studied HSE issues related to large-scale capture, transport and storage of CO_2. In their study, an almost integrated analysis has been performed; hence capture, transport and injection phases are considered in the analysis (storage phase is not included). DNV method for risk assessment of large-scale CTSC projects is SWIFT (Structured What IF Technique) analysis. SWIFT analysis is an expert panel/workshop approach to identify potential hazards and uncertainties. Prior to the workshop, a questionnaire was sent to the stakeholders in order to gather their ideas about HSE issues regarding CTSC (from capture to injection phase). The participants have mentioned the lack of an integrated approach as a concern in HSE risk management of CTSC, an approach that takes into account CTSC whole chain (Johnsen et al. 2009).

Another work on CTSC integrated risk analysis is the approach presented by Kerlero de Rosbo for the Belchatow project, in which Alstom was responsible to develop a CTSC plant for a coal-based power plant in Poland. Technical, financial, organizational, socio-political and regulatory risks associated with a large-scale CTSC project have been studied in that project. The deliverable was a risk register provided in panel discussions. (Kerlero de Rosbo 2009).

8.4 Proposed Systemic Methodology

The methodology is composed of the following steps:

1. Identifying major risks associated to CTSC (according to literature review, projects documentation and discussion with experts)
2. Assigning the risks to different CTSC subsystems and project phases
3. Defining the nature of risks and their consequences
4. Extracting the risks related to the very first phases of the project
5. Modeling the major risks using a systemic approach
6. Applying the systemic approach to model the safety control structure of different case studies

At the first step, a list of 39 major risks is made based on several references (GCCSI 2009, 2011; Longannet 2011; Feenstra et al. 2010; Kerlero de Rosbo 2009; CCP 2007; Lacq Project 2012). The risks are defined based on the project management definition of risk (*uncertain event or condition that, if it occurs, has a positive or negative effect on a project's objectives*) (PMBOK 2008).

The subsystem and project phase related to each risk are then identified. Six main phases are distinguished.

1. Opportunity:

The beginning period, when negotiations are carried out on the feasibility of CTSC project.

2. Definition and Planning:

The phase when responsibilities and authorities of stakeholders are defined, and a planning is made for the project.

3. Engineering:

Design and sizing of installations are performed in this phase.

4. Construction:

This phase deals with construction and installation of required infrastructure and equipment.

5. Operation (Injection of CO_2):

The period during which CO_2 is injected into the geological formation.

6. Post-injection (Monitoring) (also called "post-closure"):

Means the period after the closure of a storage site, including the period after the transfer of responsibility to the competent authority. (EU Directive 2009).

The third step is defining the nature of risks and related consequences based on the eight risk categories defined in "CTSC and risk management" section (Table 8.1). If we take "Project permits not obtained" as an example, the risk has a legal nature

Table 8.1 Nature of CTSC risks and their consequences

	Risk	SSubsystem[a]	Risk nature[b]	Nature of consequences[b]
1 *	Project permits not obtained	W	L	P, P/S
2 *	Technology scale-up	W	T	P, P/S, T
3 *	Public opposition	W	S	P, P/S, L
4 *	Lack of knowledge/qualified resources for operating the unit	W	T, O/H	P, P/S, HSE, O/H, T
5	Corrosion	W	T	T, P
6	Using the existing facilities (specially pipelines)	W	T	T, P
7	CO_2 out of specification	W	T	T, P, HSE
8	CO_2 plumes exceed the safe zone	W	T	P, T, HSE
9 *	Legal uncertainties	W	L	P, P/S, T, L
10	Safety related accident	W	T, O/H	T, O/H, P, HSE, S
11 *	Uncertainties in stakeholders requirements/perceptions -communication problems	W	P, P/S, HSE, O/H, T, S, L, F/E	P, P/S, HSE, O/H, T, S, L, F/E
12 *	Public availability of sensitive information	W	O/H, P/S	P, P/S, S, O/H
13 *	Change in policies/priorities	W	P/S, L	P, P/S, L
14 *	Financial crisis impact on financial support of CCS projects	W	F/E	P, P/S, F/E
15 *	Unavailability of a monetary mechanism for CO_2	W	F/E, L	P, P/S, F/E, L
16	Construction field conditions	W	T	P, T
17 *	Geographical infrastructure	W	T	T, P, P/S, S, HSE
18	Proximity to other industrial plants	W	T	T, P, HSE
19	Energy consumption	W	T	P, P/S, T
20	Maintenance and control procedures (including ESD system)	W	T, O/H	T, O/H, P, HSE
21	BLEVE	W	T	P, T, HSE
22 *	Lack of financial resources	W	F/E	P, P/S, F/E
23 *	Lack of political support	W	P/S	P, P/S, O/H, S, L, F/E
24	Phase change & material problems	W	T	P, T
25 *	High cost of project [3]	W	F/E	P, P/S, F/E
26	Lower capture efficiency due to the upstream plant flexible operation	C	T	P, T
27	CO_2 leakage from compression unit	C	T	T, P, HSE
28	Pipeline construction	T	T	P, T
29	CO_2 leakage from pipeline	T	T	T, P, HSE
30 *	Unavailability of regulations regarding different types of storage (offshore/onshore)	S	L	P, P/S, L

(continued)

Table 8.1 (continued)

	Risk	SSubsystem[a]	Risk nature[b]	Nature of consequences[b]
31	Leakage through manmade pathways such as abandoned wells	S	T	P, T
32	Well integrity	S	T	P, T
33	CO_2 migration	S	T	T, P, L, S
34	Injectivity reduction over time	S	T	P, T
35 *	Uncertainties regarding the storage performance (capacity/injectivity/containment)	S	T	P, P/S, T
36	CO_2 leakage from storage to the surface	S	T	T, P, P/S, HSE
37 *	Model and data issues	S	T	P, P/S, T
38 *	Uncertainties related to storage monitoring	S	T	P, P/S, T, HSE, S, L
39	Soil contamination	S	T	P, T

[a]W=Whole CTSC chain, C=Capture, T = Transport, S=Storage
[b]T = Technical, P=Project, S=Social, P/S=Policy/Strategy, HSE = Health, Safety, Environment, L = Legal, O/H=Organizational/Human, F/E = Financial/Economic
*Major risks affecting project progress in the first phases

and therefore, risk nature is presented by "L" (Legal) in Table 8.1. Encountering such a risk will have consequences on the project and on global and local policies and strategies regarding CTSC. Consequently, "P" (Project) and "P/S" (Policy/Strategy) are specified as nature of consequences of "Project permits not obtained". The risks are inevitably interconnected and could not be studied independently.

To analyze the reasons why a CTSC project might not be progressed, the risks related to the very first phases of the project (before engineering) are extracted from the overall list. The result is a list of 18 major risks (marked with * in Table 8.1).

Modeling safety structure of CTSC projects is carried out within the framework of the following methodology which is composed of eight steps. The steps are schematized in Fig. 8.2.

8.4.1 Define the Goal of Safety Structure

A major question about CTSC at the current stage of development is why some CTSC projects are successful to progress in particular contexts, while others fail? What are the main factors that affect the project progress?

Therefore, the goal of safety structure defined in this work is to prevent the delay or cancelation of CTSC project.

This objective could be interpreted as definition and treatment of significant risks that could prevent maintaining safety constraints.

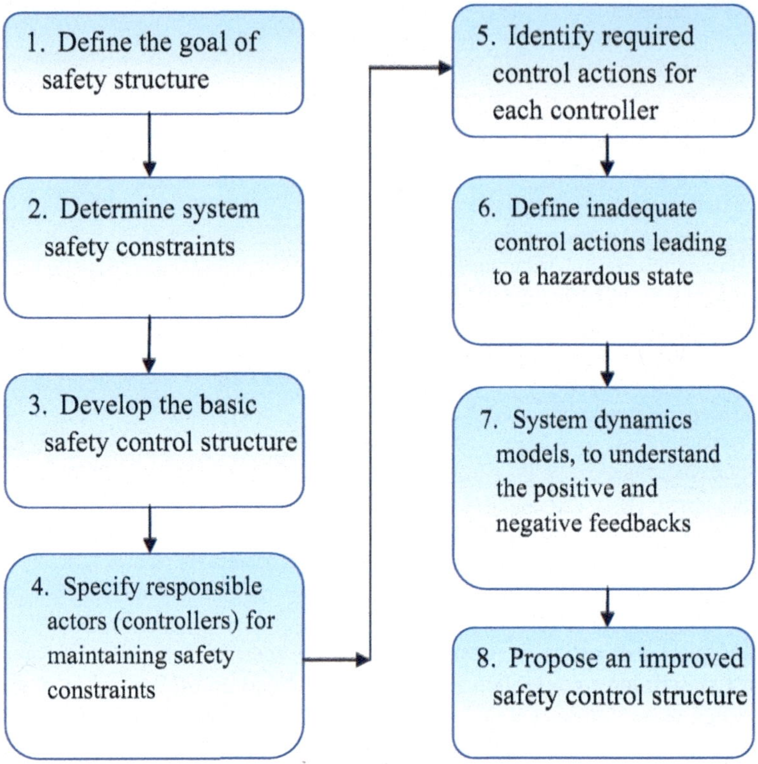

Fig. 8.2 Methodology of modeling CTSC projects safety control structure

As Leveson (1995, 2004) affirms, there are four general ways to manage risks *associated with a hazard*:

1. Eliminate the hazard from the system
2. Reduce the hazard likelihood
3. Assuring control measures when an undesired event is occurred
4. Minimize damage in case of control measures absence

8.4.2 Determine System Safety Constraints

With the goal defined in the first step, the following constraints could be fixed for the system:

1st system safety constraint: The project must not be delayed or cancelled.
2nd system safety constraint: Measures of control must be provided in case of delay or cancellation.

8.4.3 Develop the Basic Safety Control Structure

The structure for CTSC is context specific, depending on several factors including location, population density and historic issues (CCP 2012). However, the following stakeholders are present in almost all cases:

- Project owner
- Politicians and Policy makers (National and Local)
- Regulators
- External experts
- Local population
- NGOs
- Media

Each stakeholder is a "controller" of the system, who is responsible for maintaining specified safety constraints.

8.4.4 Specify Responsible Actors for Maintaining Safety Constraints

For the safety constraints introduced in the second step, project owner is directly responsible. In fact, project owner is the endogenous controller, while other actors are exogenous controllers who could affect the system and decisions of the project owner.

8.4.5 Identify Required Control Actions for Each Controller

Required control actions are the tasks that should be performed in order to maintain the safety constraints. These actions are risk specific.

8.4.6 Define Inadequate Control Actions Leading to a Hazardous State

Hazardous state is a state *that violates the safety constraints* (Leveson 2004).

Leveson presents four general types of inadequate control:

1. *A required control action is not provided.*
2. *An incorrect or unsafe control action is provided.*

3. *A potentially correct or adequate control action is provided too late (at the wrong time).*
4. *A correct control action is stopped too soon.*

8.4.7 Develop System Dynamics Models, and Especially Causal Graphs

The purpose is to study the positive and negative feedback loops which are involved in the process of maintaining safety constraints. From a system dynamics point of view, positive feedback loops are the ones which *destabilize the system and cause them to run away from their current situation* whilst negative feedback loops stabilize the system (Radzicki and Taylor 1997). In a positive feedback loop, the variables change in the same direction, whereas they will change in the opposite direction in a negative feedback loop.

8.4.8 Propose an Improved Safety Control Structure

The improved structure is developed based on the analysis of inadequate control actions and causal graphs. Causal graphs are the diagrams containing the network of feedback loops or the interactions of system variables.

Different types of risk and several case studies have been modeled using this approach. Examples of modeling are explained here-after.

8.5 Risk of Technology Scale-Up Problems

As Herzog argues, in order to realize the objective of cutting greenhouse gas emissions by 2050, we will need to capture and store gigatonnes of CO_2 every year. Following challenges are mentioned for CTSC scale-up (Herzog 2009):

- Cost
- Infrastructure
- Subsurface Uncertainty (Capacity & Long-term Integrity)
- Regulatory Framework
- Long-term Liability
- Public Acceptance

The safety constraint for the current example is as follows:

Safety constraint: Measures must be put in place to avoid the risk of delay or cancellation due to technology scale-up issues.

Herzog argues that CTSC technology scale-up is a complex issue that depends on various factors. Hence, several stakeholders are involved as the controllers of such a complex subject.

Project owner and Competent Authorities could be considered as endogenous controllers, while Regulators, Local population and others are the exogenous controllers, which have impacts on the decisions of endogenous actors (Fig. 8.3).

The analysis of technology scale-up risk could be summarized as shown in Table 8.2:

The feedbacks involved in controlling the rate of increase of technology scale-up problems are illustrated in the following causal graph:

Risk of technology scale-up issues and the rate of risk increase are presented as stock and flow variables respectively. There are three types of variables in system dynamics modeling: Stock, Flow, and Auxiliary (Control) variables.

- Stock (or Level): is an accumulation within the system, for example inventories, goods in transit, bank balances, factory space and the number of employees (Forrester 1968). The quantity of stock is the integral of difference between its outflow and inflow.
- Flow (or Flow Rate or Rate): is another element of a model structure that defines the flows between the stocks in the system. (Forrester 1968)

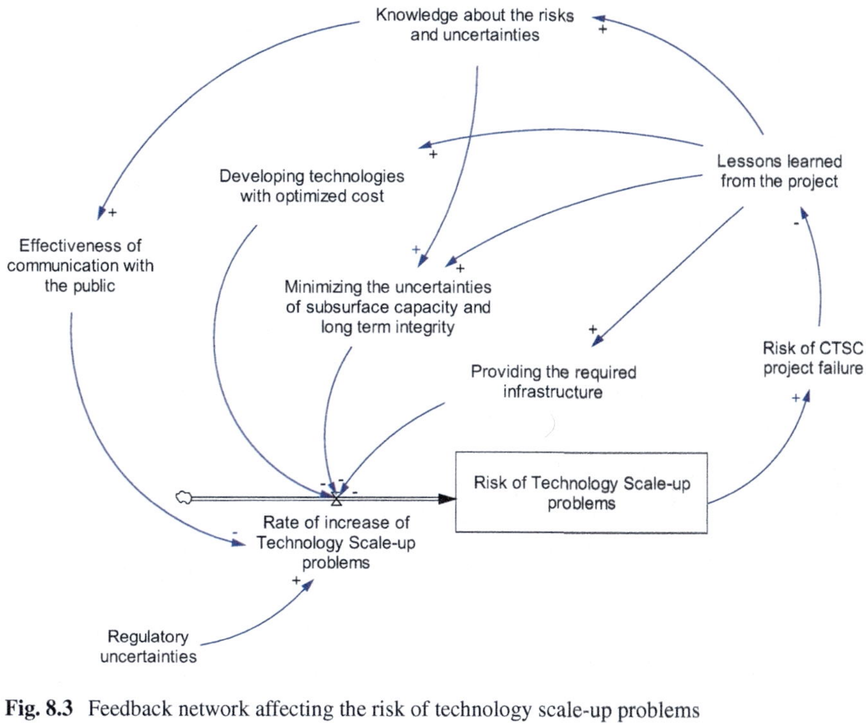

Fig. 8.3 Feedback network affecting the risk of technology scale-up problems

Table 8.2 Summary of second example, risk of technology scale-up problems

Risk: technology scale-up problems
Safety constraint:
Measures must be put in place to avoid the risk of delay or cancellation due to technology scale-up issues
Who is responsible for maintaining the safety constraint?
Project owner & competent authorities
Required control actions:
For project owner: Providing the required elements to minimize the uncertainties associated to subsurface and storage long term liability
Ensuring the public acceptance of the project development
Providing the necessary infrastructure
Providing technologies with optimized acceptable costs
And for competent authorities: Providing regulatory frameworks for CTSC activities
(Examples of) Inadequate control actions leading to a hazardous state:
Acceptable costs are not proposed for CTSC technologies Best available costs are proposed by the developers, but not accepted by the authorities because of uncertainties/ambiguities of acceptable costs range Economically acceptable options are made available too late Research and developments are not continued to find more economic technological possibilities (as soon as an option is accepted by the authorities) (there is a risk of changing range of acceptable costs over time)

- The following equation shows the mathematical relation between the stocks and flows:

$$levels_t = \int_0^T rates_t dt \quad \text{or} \quad \frac{d}{dt} levels_t = rates_t$$

- Auxiliary (Control) variable: is a variable that is *computed from other variables at a given time. Auxiliaries are typically the most numerous variable type.* (Vensim Reference Manual 2010)

Challenges of scale-up, introduced at the beginning of this section, are the control variables affecting the rate of increase of technology scale-up problems. The challenges are represented as "*Providing the required infrastructure*", "*Minimizing the uncertainties of subsurface capacity and long term integrity*", "*Developing technologies with optimized cost*", "*Effectiveness of communication with the public" and "Regulatory uncertainties*".

As discussed earlier, CTSC is a novel complex technology for which risks cannot be analyzed and managed separately. The interrelations of risks create a context which has the potential to give rise to a hazardous state. Therefore, the interconnections of risks shall be modeled and studied. Inter-relations of the major risks are

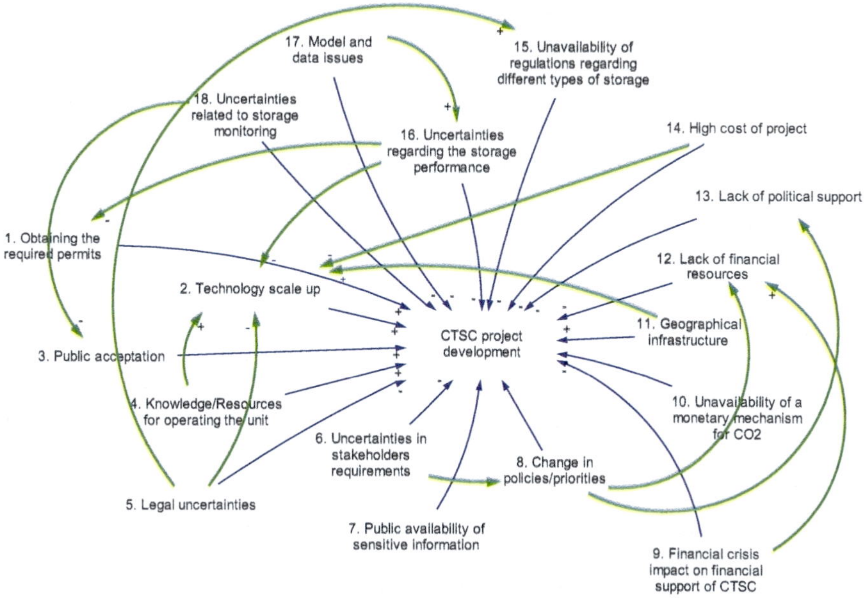

Fig. 8.4 Interconnections of major risks affecting CTSC projects progress

illustrated in the following causal graph. The green bold feedbacks represent the risks interconnections (Fig. 8.4).

"*Technology scale up*", for instance, is influenced by five other risks: "*Knowledge/Resources for operating the unit*", "*Legal uncertainties*", "*Geographical infrastructure*", "*High cost of project*" and "*Uncertainties regarding the storage performance*".

Three case studies are selected for modeling safety control structure. The selection is based on project success level.

The first case is Barendrecht, in the Netherlands, which was cancelled due to public opposition and lack of local support.

The second example is Lacq, as the first CTSC pilot plant in France, in which CO_2 injection is going on in spite of some technical challenges.

The third case is Weyburn, as a successful industrial scale EOR project in the North America, which has to deal with some questions.

Barendrecht example is presented and discussed here after.

8.6 Modeling Barendrecht Case Study

Barendrecht was a CTSC integrated project, planned to inject 400,000 tonnes CO_2 per year. CO_2 was produced in a hydrogen production plant and planned to be injected in two depleted gas fields. The capture plant is located about 20 km from

Barendrecht, a town located in the west of the Netherlands. Barendrecht is situated at around 14 km of Rotterdam, the second largest city in the Netherlands. The population of the city is about 44,000 people. A pipeline of 16.5 km was designed to transport the captured CO_2 to the storage location. The first gas field (Barendrecht) could store about 0.8 million tonnes of CO_2 at a depth of 1700 m. The second gas field (Barendrecht-Ziedewij) could store about 9.5 million tonnes of CO_2 at a depth of 2700 m.

Shell was the owner of the project, and a financial support of 30 million euros was invested by the government for this project. Shell would also have the benefit of emission saving under ETS (Emissions Trading System) program.

The tender was announced by the Dutch government in 2007. In early 2008, Shell was selected as the winner of the tender. Debates have begun from then on, when the project was presented to local community. The first phase of injection was planned to start in 2011 for a duration of 3 years. Injection in the second gas field was planned to begin in 2015 for 25 years (Feenstra et al. 2010).

In November 4th, 2010, Dutch Minister of Economic Affairs, Agriculture and Innovation announced that the project is cancelled. *The delay of the CO_2 storage project for more than 3 years and the complete lack of local support are the main reasons to stop*. However, the minister believes that Barendrecht experiences are valuable for further development of CO_2 storage in the Netherlands. So, *Barendrecht cancellation does not mean the end of CO_2 storage in the Netherlands*. (CCJ 2010; Netherlands Government 2010).

The application of the methodology for Barendrecht project is discussed in next paragraphs. The purpose is to understand the weaknesses of the project safety structure, and the points that could be improved to avoid the delay and stop.

The main stakeholders and their responsibilities are as following (Feenstra et al. 2010):

- National government: was engaged via two ministers: Ministry of Economic Affairs (EZ) and Ministry of Housing, Spatial Planning and Environment (VROM).
- EZ established a group (Task force CCS), with representatives of industry, NGOs and local governments, to support CTSC development in the Netherlands.
- Local governments: were involved at two levels: provincial and municipal.
- The executive board of the provincial government was responsible for environmental permitting procedures. An environmental protection agency (DCMR) was appointed by the provincial deputy to execute the leadership of a consultation group (BCO_2). BCO_2 was the administrative consultation group of Barandrecht project.
- At the municipal level, governments of Barendrecht and Albrandswaard were involved. Albrandswaard population did not raise many concerns about the project, probably because a few numbers of their houses were located directly above the gas fields. Barendrecht government was more actively involved.
- Project developers: Three companies were engaged. Shell was the initiator and responsible for storage and monitoring. Two other companies were collaborating

with Shell for capture and transport. NAM (Netherlandse Aardolie Maatschappij BV), the Netherlands biggest oil and natural gas producer, was *responsible for existing natural gas production from the gas fields in Barendrecht*. OCAP (Organic CO_2 for Assimilation of Plants) was responsible for CO_2 transport.

- External experts, consultants and research organizations: Several external experts were involved, mainly by project developers, for environmental studies of CO_2 storage and to answer the questions from municipality in the public meetings.
- NGOs: Several NGOs were also active for or against the project. Greenpeace is opposed to CTSC, at national and international scales. Uncertainties about subsurface capacity to store CO_2, energy waste, risk of CO_2 leakage and expensiveness are the principal concerns of Greenpeace regarding CTSC technology (Rochon et al. 2008). SNM, the Netherlands Society for Nature and Environment, believes that CTSC is essential as an intermediate step towards clean energy.
- Local population: The people who live in the neighborhood of CO_2 storage location are significant stakeholders of CTSC projects. In Barendrecht case, they were represented by the municipal government.
- Media: Local and national newspapers, as well as televisions, websites and magazines were another actors who were involved in distributing information on the project.

The organizational structure of Barendrecht project is illustrated in Fig. 8.5.

Lessons learned from the project confirm that communication problems are the main issues resulted in the opposition to the project. The most significant subjects affecting the effectiveness of the safety control structure are as following:

1. As showed in Fig. 8.5, there is no connection between the national and local governments. The lack of such connection reinforced the public opposition.
2. Delays in some required actions made the community resist to the project. Some examples are presented in Fig. 8.5. Establishment of the administrative consultation group (BCO_2) by the national government occurred rather belatedly, after the start of local opposition. Delay symbol on the connection between National Government and BCO_2 illustrates such late reaction. In addition, presentation of the project to the community (Local Governments and population) happened with delay. Some information on the project was not communicated upon request, especially due to confidentiality issues.
3. Regulatory responsibilities were not so clear in the project context. Changing the project regulatory framework was another reason for which the opposition occurred. In the new framework, the project would be considered as a one having national impacts. Therefore, National Government was authorized for all needed permissions, even those normally awarded by local governments.
4. Another issue is the lack of mutual connection between the stakeholders in some cases. For example feedbacks of local governments were not taken into consideration by the project developers, although the project had been presented to the local community.

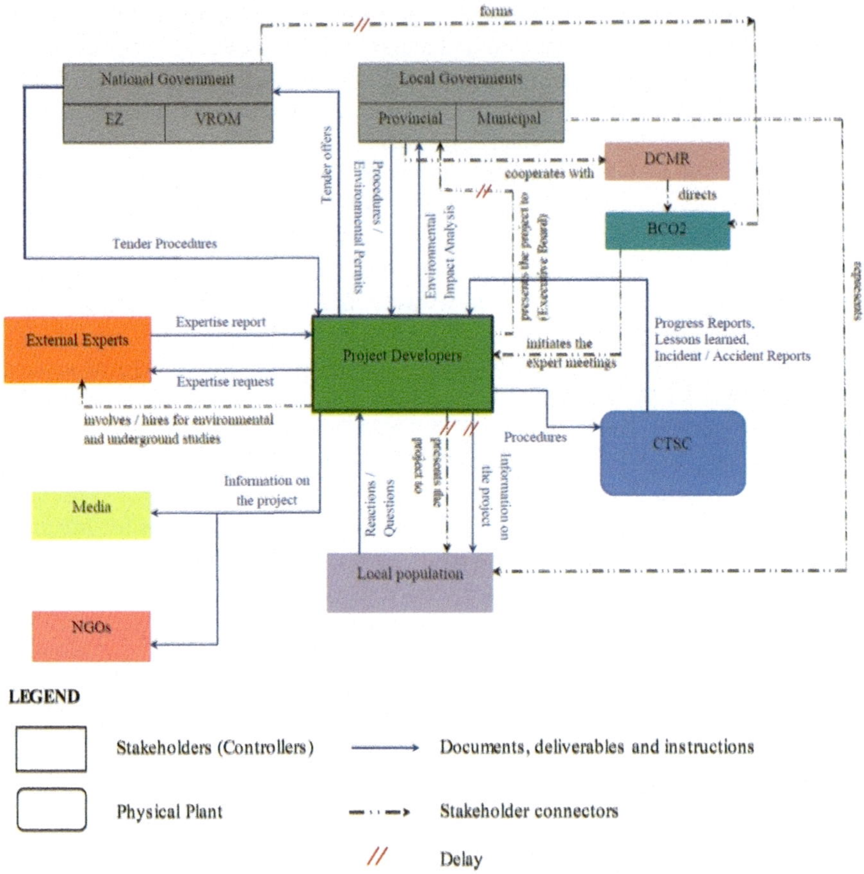

Fig. 8.5 Barendrecht safety control structure, initial model
Rectangles with sharp corners symbolize the stakeholders (controllers), while the round-corner rectangle (CTSC) stands for the physical plant.
Dash lines are the stakeholders connectors, which show the relations of actors. A number of generic connector types have been proposed by (Dulac 2007) and (Stringfellow 2010). Documents, deliverables and instructions exchanged between the actors are represented by solid lines.
(//) on the arrows represents delay, which is also a system dynamics concept.
When delay symbol (//) is put on a connection, it means that the action is carried out with delay.

In some cases, mutual connections are not available for a particular reason. For instance, NGOs preferred to announce their opinion in the national level, instead of on this specific project. Therefore, no feedback is considered from NGOs to the project developers. The media also tried not to influence opinions. Thus, no direct connection is available from the media to the project developers.

An improved version of the project safety control structure (based on the 1st, 2nd and 4th abovementioned issues) is presented in Fig. 8.6. The added elements are presented in orange. Delays existed in Fig. 8.5 are removed in the proposed improved model (Fig. 8.6).

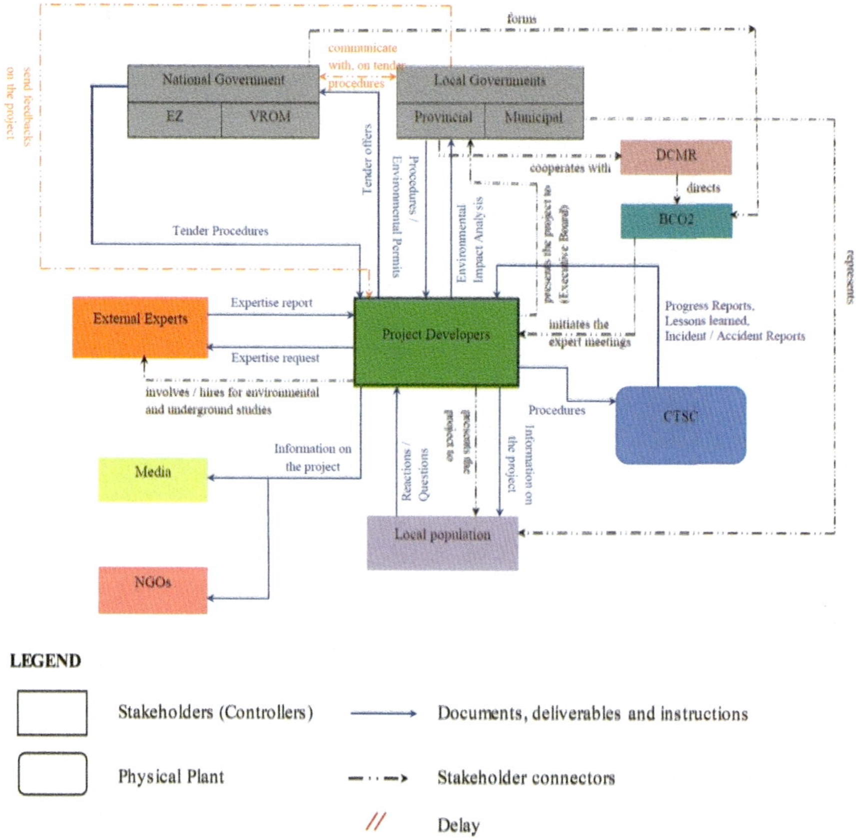

Fig. 8.6 Barendrecht safety control structure, improved model

The areas of concern in Barendrecht safety control structure could be summarized as inadequate control actions led to the project failure (Table 8.3).

Barendrecht example confirms that all potential interconnections are not identified in the risk network (Fig. 8.7). Lessons learned from the project assert that legal uncertainties/modifications, uncertainties in stakeholders requirements and lack of political support could lead to public opposition. Risk interconnections are shown in Figs. 8.7 and 8.8.

In order to propose an improved control structure for CTSC projects, the three case studies are compared from context and risk point of view. Lessons learned from further development of projects will provide useful information to improve and complete this analysis. The proposed structure (Fig. 8.9) confirms the importance of communication among stakeholders. *An iterative and interactive dialogue between the key stakeholders is also recommended* by (Koornneef et al. 2012) *to ensure that state of the art knowledge is included in the risk management of storage projects.*

Table 8.3 Principal Inadequate Control Actions leading to Barendrecht project failure

Principal inadequate control actions leading to Barendrecht project failure:
Delay in presenting the project to the community
Avoiding to share certain (confidential) information
Delay in providing feedback on stakeholders concerns or questions
Unavailability of communication between some stakeholders, such as national and local governments
Unavailability of a correctly specified regulatory framework

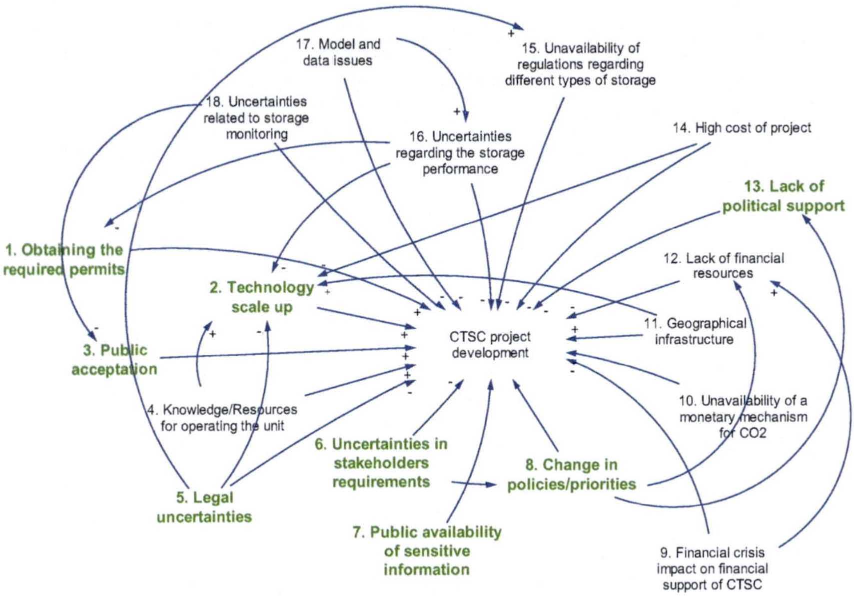

Fig. 8.7 Interconnections of major risks affecting Barendrecht project progress

In Fig. 8.9, solid lines represent documents and information exchanged between the stakeholders, not necessarily in a dynamic manner. Dash lines show the flow of dynamic interchange, i.e. what should be maintained throughout the project life.

Global, National and Local Governments are regrouped in a box, since the relationship of other stakeholders with the governments is varied in different regions.

Regulators are asked by Policy Makers for regulatory frameworks. Global policies and permitting procedures are defined (by Global Policy Makers) for CTSC according to regulatory frameworks and climate change policies. National policies and permitting procedures are transposed to national contexts by National Policy Makers, who shall be continuously in communication with Global and Local Policy Makers.

Tender procedures are sent to the Project Owner by the government. The Project Owner returns the tender offers and if the offer is accepted, project permits will be provided in reply to the authorization request of the Project Owner.

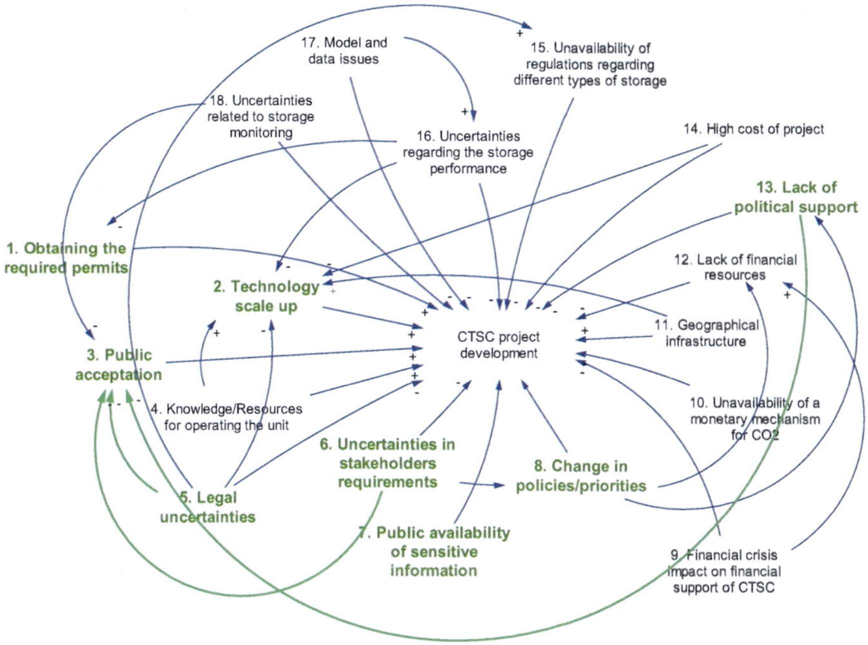

Fig. 8.8 Interconnections of major risks affecting Barendrecht project progress, modified according to lessons learned

The hatched squares (including Governments, Project Owner and External Investors) represent potential investors of the project who should intercommunicate on the funds allocated for the project. There are still several uncertainties about the actors who have to pay for developing CTSC technologies.

External Experts are always engaged to provide expertise usually on technical aspects of the project.

Information on the project has to be shared dynamically with all stakeholders including Local Population, NGOs and Media.

Communication is also essential between governments, NGOs and Local Population, since local communities need to be assured of political support of their policy makers in order to accept CTSC as a novel beneficial technology.

Delays, especially in communication, have to be minimized.

Figure 8.9 underlines the significance of information feedback loops within the safety control structure of CTSC projects. Information feedbacks allow the actors to improve their mental models, decisions, strategies and decision rules.

Dulac asserts this opinion by remarking that *improving mental models will consequently improve the quality of safety-related decision-making … and the performance of organizations and systems* (Dulac 2007; Leveson 2009). Mental models are the schemas of human beings which help them make decisions. Investigating mental models of both experts and lay people provide essential information for communication. (Skarlatidou et al. 2012).

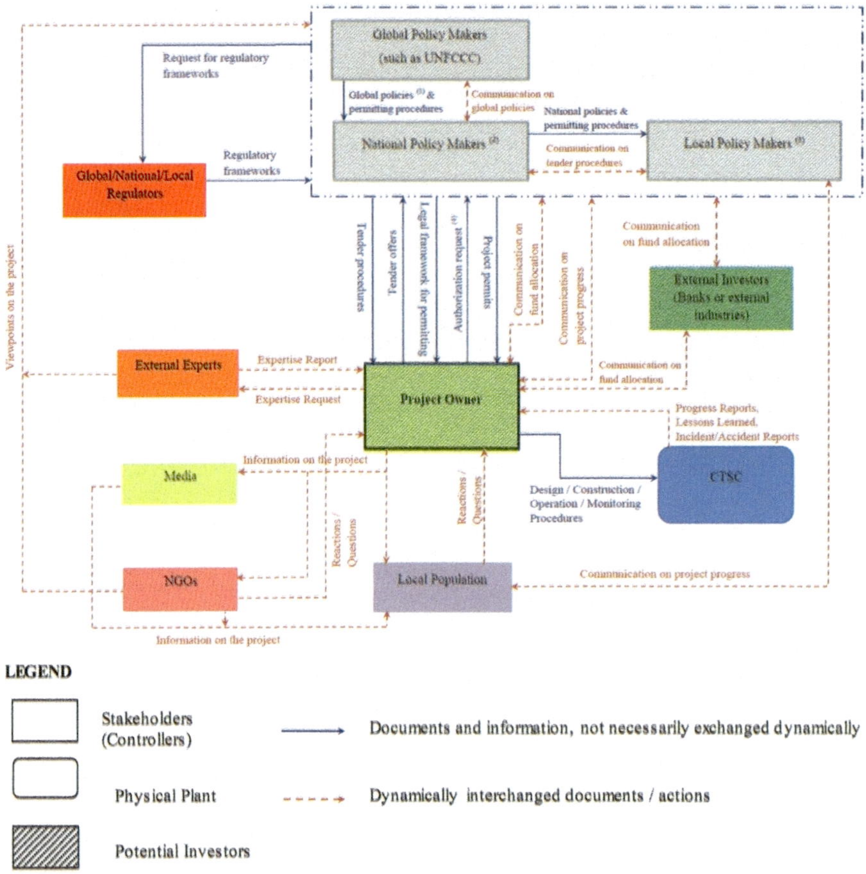

Fig. 8.9 Proposed safety control structure for CTSC projects
(1) Global policies according to regulatory frameworks
(2) Including Policy Makers in the scale of zones (EU, US, etc.) and countries
(3) Including Policy Makers in the scale of regions and communities
(4) Including EIA, Hazard Analysis and HSE concerns

A systemic risk management framework for CTSC projects has been proposed in this work. The approach is founded on the concepts of systems thinking, STAMP, STPA and system dynamics. The objective is to provide a means of decision making for CTSC projects development in the actual context where the future of the technology is uncertain. Risk management is considered as a means of control that should be able to propose a control structure for the whole system. Stakeholders are viewed as controllers of the system.

Major risks involved in CTSC projects progress have been categorized in eight groups including Technical, HSE (Health, Safety and Environment), Policy/Strategy, Legal, Organizational/Human, Financial/Economic, Social and risks concerning the Project.

Thirty nine risks have been identified according to literature review, available projects information and discussions with experts. The risks have been classified for Capture, Transport and Storage subsystems and for different phases of the project. Opportunity, Definition and planning, Engineering, Construction, Operation (Injection of CO2) and Post-injection (Monitoring) are the main project phases that have been taken into account. In order to analyze the risks preventing project progress, the ones related to the phases prior to engineering have been selected and modeled by the proposed methodology. The aim was to study the feedback networks affecting the risks amplification. The analysis has been started from stock / flow models of each risk. Models have been subsequently grouped together in order to study interconnections of risks and feedback loops result in project failure or success.

The advantages of risk management methodology in this work compared to the existing ones such as (GCCSI 2009) and (Kerlero de Rosbo 2009) can be summarized in three points:

1. Presenting more comprehensive list and categories of risks related to CTSC chain
2. Taking into account the complex network of risk interconnections by proposing a systemic modeling framework
3. Underlining the significance of stakeholders role in the project success or failure, by proposing a modeling approach for safety control structure of projects and analyzing required and (potential) inadequate control actions of stakeholders in relation to each risk

Development of CTSC projects will provide lessons learned for improving the models. New information could be used to put figures on the variables of models in group modeling panels. Figures help to make semi-quantitative analyses which may be more understandable for some stakeholders. Effects of feedback loop networks on the probability and intensity of risks could also be studied in semi-quantitative approaches.

It is suggested to apply the methodology for a CTSC project in feasibility study or definition phase. This application will provide a decision making support for the development of projects.

References

Carayon P (2006) Human factors of complex sociotechnical systems. Appl Ergon 37(2006):525–535. https://doi.org/10.1016/j.apergo.2006.04.011

CCJ (2010) Shell Barendrecht project cancelled, Carbon Capture Journal, November 5, 2010. Accessed 22 June 2012. http://www.carboncapturejournal.com/displaynews.php?NewsID=676

CCP (2007) Public perception of carbon dioxide capture and storage: prioritized assessment of issues and concerns. Summary for policy-makers, CO2 capture Project®, March 2007

CCP (2012) CCS stakeholder issues, Review and Analysis, CO2 Capture Project®, Final Report, February 2012

De Coninck H, Flach T, Curnow P, Richardson P, Anderson J, Shackley S, Sigurthorsson G, Reiner D (2009) The acceptability of CO_2 capture and storage (CCS) in Europe: An assessment of the key determining factors, Part 1. Scientific, technical and economic dimensions. Int J Greenhouse Gas Control 3(2009):333–343. https://doi.org/10.1016/j.ijggc.2008.07.009

Dulac N (2007) A framework for dynamic safety and risk management modeling in complex engineering systems. MIT, Cambridge, MA

EU Directive (2009) Directive 2009/31/EC of the European Parliament and of the Council of 23 April 2009, on the geological storage of carbon dioxide and amending Council Directive 85/337/EEC, European Parliament and Council Directives 2000/60/EC, 2001/80/EC, 2004/35/EC, 2006/12/EC, 2008/1/EC and Regulation (EC) No 1013/2006

Farret R, Gombert P, Lahaie F, Roux P (2009), Vers une méthode d'analyse des risques globale de la filière CSC, intégrant plusieurs échelles de temps, Rapport Scientifique, INERIS, 2008-2009, pp 100–103

Feenstra CFJ, Mikunda T, Brunsting S (2010) What happened in Barendrecht?, Case study on the planned onshore carbon dioxide storage in Barendrecht, the Netherlands, ECN/CAESAR, November 3, 2010

Forrester JW (1968) Industrial dynamics. The MIT Press, 5th printing, May 1968, Cambridge, MA

GCCSI (2009) Strategic analysis of the global status of carbon capture and storage, report 5: synthesis report, Global CCS Institute, Canberra, Australia

GCCSI (2011) The global status of CCS: 2011, Global CCS Institute 2011, Canberra, Australia, ISBN 978-0-9871863-0-0

GCCSI (2014) The global status of CCS: February 2014, Global CCS Institute 2014, Australia

Gerstenberger M, Nicol A, Stenhouse M, Berryman K, Stirling M, Webb T, Smith W (2009) Modularised logic tree risk assessment method for carbon capture and storage projects. Energy Procedia 1:2495–2502

Hellström T (2009) New vistas for technology and risk assessment? The OECD programme on emerging systemic risks and beyond. Technol Soc 31(2009):325–331. https://doi.org/10.1016/j.techsoc.2009.06.002

Herzog H (2009) Scaling-up CCS: from megatonnes to gigatonnes, presentation in Copenhagen, March 11, 2009

IEA (2010) Energy technology perspectives 2010, scenarios and strategies to 2050. International Energy Agency, OECD/IEA, Paris

IRGC (2010) The emergence of risks: contributing factors, report of international risk governance Council, Geneva, 2010, ISBN 978-2-9700672-7-6

Johnsen K, Holt H, Helle K, Sollie OK (2009) Mapping of potential HSE issues related to large-scale capture, transport and storage of CO_2, DNV Report for Petroleumstilsynet

Kerlero de Rosbo G (2009) Integrated Risk Analysis for large-scale CCS projects implementation, Thèse professionnelle du Master Spécialisé en Ingénierie et gestion de l'Environnement (ISIGE)

Koornneef J, Ramirez A, Turkenburg W, Faaij A (2012) The environmental impact and risk assessment of CO_2 capture, transport and storage - An evaluation of the knowledge base. Prog Energy Combust Sci 38(2012):62–86. https://doi.org/10.1016/j.pecs.2011.05.002

Lacq Project (2012) Documents concerning Lacq Project, available on the regional prefecture website, Accessed from November 2010 to October 2012. http://www.pyrenees-atlantiques.pref.gouv.fr/Politiques-publiques/Environnementrisques-naturels-et-technologiques/Pilote-d-injection-de-CO2/Projet-de-captage-CO2-de-Total/

Leveson N (1995) Safeware, System Safety and Computers, Addison-Wesley Publishing Company, ISBN 0201119722

Leveson N (2004) Model-based analysis of socio-technical risk, Massachusetts Institute of Technology, Engineering Systems Division, Working Paper Series, ESDWP- 2004-08, December 2004

Leveson N (2009) Engineering a safer world, system safety for the 21st century, Massachusetts Institute of Technology, July 2009 (a draft book), Accessed 1 Dec 2011. http://sunnyday.mit.edu/safer-world.pdf

Longannet (2011) FEED documents of Longannet project, published online http://www.decc.gov.uk/en/content/cms/emissions/ccs/demo_prog/feed/scottish_power/scottish_power.aspx

Netherlands Government (2010) http://www.rijksoverheid.nl/nieuws/2010/11/04/co2-opslagproject-barendrechtvan-de-baan.html

OECD (2003) Emerging risks in the 21st century, an agenda for action, organization for economic co-operation and development. OECD Publications Service, Paris

PMBOK (2008), A guide to the Project Management Body of Knowledge (PMBOK® Guide), ISBN 978-1-933890- 51-7, 4th edition, Project Management Institute, USA

Radzicki MJ, Taylor RA (1997) Introduction to system dynamics, US Department of Energy http://www.systemdynamics.org/DL-IntroSysDyn/inside.htm

Rochon E, Kuper J, Bjureby E, Johnston P, Oakley R, Santillo D, Schulz N, Von Goerne G (2008) False Hope: Why carbon capture and storage won't save the climate, Greenpeace, Published in May 2008 by Greenpeace International, Amsterdam, The Netherlands

Skarlatidou A, Cheng T, Haklay M (2012) What do lay people want to know about the disposal of nuclear waste? A mental model approach to the design and development of an online risk communication. Risk Anal 32(9):2012. https://doi.org/10.1111/j.1539-6924.2011.01773.x

Stringfellow MV (2010) Accident analysis and hazard analysis for human and organization factors, PhD thesis submitted to Massachusetts Institute of Technology, Department of Aeronautical and Astronautical Engineering, October 2010

UNFCCC (2012) Report of the conference of the parties on its seventeenth session, held in Durban from 28 November to 11 December 2011, Addendum, part two: action taken by the conference of the parties at its seventeenth session, contents: decisions adopted by the conference of the Parties, 15 March 2012. Accessed 13 Oct 2012. http://unfccc.int/resource/docs/2011/cop17/eng/09a01.pdf

US-EPA (2010) Human-related sources and sinks of carbon dioxide, United States Environmental Protection Agency. Accessed August 2010 http://www.epa.gov/climatechange/emissions/co2_human.html#industrial

Vensim Reference Manual (2010) Copyright 2010, Ventana Systems, Inc. Revision data August 22, 2010. Accessed February 2011 http://www.vensim.com/

General Conclusion

Designing and developing a systemic safety model requires the acquisition of background knowledge related to control and systems theories. Thus, the concepts of systemic, control, control loop, feedback, etc. must be fully understood, which in turn requires a good understanding of the concepts belonging as much to the 'systemic' domain as the safety domain. This dual understanding of the 'foundations' raises many questions:

- What is a system? How should it be understood?
- What is complexity?
- What are the different types of systems?
- What is a dynamic system? What is feedback?
- What is dynamic behaviour?
- What is safety? What is a safety system process?
- What is a hazard analysis technique?
- How can analysis techniques be integrated into the life cycle of a system?
- What is a hazard? What is an accident?
- What is the triptych that makes up a hazard analysis approach?
- Etc.

However, a good knowledge of the safety and systemic domains is insufficient to exploit systemic accident models: the latter require that their user has an intimate understanding of their challenges, which goes far beyond applying safety principles in a given system. It is rather a matter of understanding a system in which safety is an emerging property, and therefore to highlight a complementarity between a level of safety and a systemic optimization.

Nowadays, it is accepted that systemic accident models can overcome the limitations of traditional models based on the analytic approach, which has proved to be

F. Guarnieri, E. Garbolino (eds.), *Safety Dynamics*, Advanced Sciences and Technologies for Security Applications,
https://doi.org/10.1007/978-3-319-96259-7

inadequate for dynamic systems. The first, so-called linear, accident models were only able to provide 'stable' solutions and predict four types of unstable situations:

- stable oscillations, i.e. slight variations of the system, that can be brought back into balance by linear and proportionate responses;
- exponential growth oscillations, i.e. systemic behaviour characterized by positive and growing feedback loops, and negative loops that can 'cushion' variation;
- asymptotic stability, characterized by negative feedback loops that make it possible to achieve a goal or stabilize a system;
- unlimited growth, characterized by positive feedback loops leading to exponential growth.

One of the advantages of linear accident models is their ability to provide accurate predictions about the future state of a system, while systemic accident models can only provide one law governing the evolution of a system. However, the dynamics of system, and sociotechnical systems are nonlinear. To understand an 'accident' system, in the context of a linear accident model, it is sufficient to model it by examining its behaviour when it is at steady state. Once this modelling is done, it is then possible to predict the general behaviour of the system when the initial conditions are known. However, these linear accident models are only valid and really appropriate when the system is in a situation that is close to a stable equilibrium.

Systemic accident models, based on general system theory and control theory, focus on the analysis of a system that is in a steady state and evolving towards an accidental state. As soon as information concerning the elements of the system is known, it becomes possible to describe its dynamic behaviour, to analyse it, or to carry out an accident investigation. In other words, as soon as the initial conditions of a dynamic system are known, it is possible to describe its trajectory and the law governing its evolution.

Systemic accident models are thus able to produce types of solutions that are different to those described by traditional accident models, while leaving room for structural changes. They also make it possible to propose ways to reinforce the robustness of a system, i.e. its sensitivity to small disturbances in order to keep it in a 'safe' state.

In this context, a systemic accident model, such as the STAMP model, seeks to reinforce the robustness and stability of a system in the face of disturbances. Two types of stability exist in a dynamic system: dynamic and structural. Dynamic stability is translated by a system behaviour in the face of a disturbance that is exterior to the system (for example, an external 'noise'). In this context, the system will seek to provide a response that is adapted to an external signal while maintaining a 'safe' state. Structural stability (which is at the heart of the STAMP model, with the establishment and study of a hierarchical structure) concerns the control of the structure (composed of a large number of elements in nonlinear interaction). This structural stability remains internal to the system and the maintenance of a 'safe' state depends on the capacity of the system to maintain internal control over its structure.

Systemic accident models have therefore been developed to take into account the complexity of sociotechnical systems. In much of the literature, complexity is synonymous with a large number of interactions between elements. This condition is necessary but not sufficient. In practice, a system can be composed of a large number of interactions, but lack any sign of complexity (for example, an ice cube, which is composed of a large number of molecules in linear interaction, is in a state of stable equilibrium). The additional condition for characterizing complexity is the notion of nonlinearity. In this case, a system composed of many elements in nonlinear interaction can be considered as a complex system. This is the case for sociotechnical systems.

As has been discussed throughout the book, systemic accident models are based on Bertalanffy's general system theory.[1] In this context, they study systems that are in (or close to) a state of stable equilibrium (when they are subject to disturbances). However, complex systems are never in a state of stable equilibrium, and it is therefore not particularly useful to attempt to study them as complex systems with tools that are based on a theory that studies systems that are in (or close to) a stable equilibrium.

This contradiction underlies the STAMP accident model.

The STAMP accident model is based on two theories: system theory put forward by Ludwig von Bertalanffy, and control theory, notably the work of Norbert Weiner on the concept of feedback. These theories set out the characteristics of systems described in the framework of classical science—the notions of equilibrium, linear causality and negative feedback. These notions make it possible to explain the context in which accident models have developed.

The notion of equilibrium corresponds to the state of a system when it automatically corrects deviations from its trajectory determined by basic laws. In this context, a sociotechnical system is a set of systems in which the natural state is a stable equilibrium where any deviation is attenuated or corrected by negative feedback. When a disturbance is too great and does not allow the return to the initial equilibrium state, the system can migrate to a new state of equilibrium.

Linear causality postulates that there is a direct link between the cause acting on a system and the changes in the structure of the system, in other words, cause and effect are proportional.

Finally, the notion of negative feedback, found in the work of Newton, Darwin or Bertalanffy translates a mechanism that exists when a system departs from its equilibrium, due to external disturbances or fluctuations that are inherent in its dynamic, and which attenuates their effects to allow the system to return to its initial equilibrium state. The notion of negative feedback has largely inspired the field of cybernetics.

Accident models are therefore developed to describe, interpret and predict an accident. Here, the accident model approach is based on a deterministic approach, which constitutes the basic principle of any prediction. In practice, the fundamental objective of an accident model is predictive; it focuses on the analysis of hazards in

[1]Please see : Von Bertalanffy, L. (1968). General system theory. *New York*, *41973*(1968), 40.

a system (static or dynamic, linear or nonlinear) in order to prevent behaviours that can migrate a 'safe' system to an accidental state. All accident models have been designed with the aim of understanding and prediction in different types of systems, such as physical or sociotechnical systems. Linear accident models are limited, and represent an approximation of reality that is only useful when the system is close to a stable equilibrium state. However, an accidental phenomenon never occurs when a system is close to a stable equilibrium. These accident models have been built on a deterministic process, combining an analytical and a systemic approach (now considered complementary). They have always assumed that the studied systems are 'integrable' (linear) and deterministic. However, systems such as sociotechnical systems integrate many elements that interact nonlinearly, generating many types of behaviours and therefore changes.

A consequence of this is that systemic accident models take account of the notion of 'time'. At present, the modelling of sociotechnical systems is based on the theory of integrable dynamic systems, in other words, systems that are close to a stable equilibrium state, in which (ongoing and never-ending) time flows. In these systemic accident models, the system is dynamic and evolves over time. However, a systemic accident model that studies a dynamic system in stable equilibrium state cannot be made to evolve continuously, because any response to a disturbance can only be 'linear'. A system, whatever it may be, in a state of stable equilibrium cannot evolve, because its trajectory is a fixed point that does not move. A stable equilibrium state is only governed by linear processes and does not drive the system to change. A system only evolves in a state of instability based on irreversible processes. It then becomes difficult to speak of dynamics and evolution over time, as defined in the framework of Bertalanffy's general system theory, in the context of systems in an unstable equilibrium—as is the case for sociotechnical systems.

Time is a parameter that imposes itself on any system. It must be considered as a degree of freedom, or even as an emergent property, and not as a universal parameter imposed by default on any system. The physical laws that general system theory is based on do not distinguish between the future and past directions of the 'arrow of time' in a non-integrable dynamic system. However, this notion of the arrow of time makes perfect sense in the theory of non-integrable dynamic systems.

Thus, time (notably, the arrow of time[2]) emerges from the instability and irreversibility of processes. Entropy is born out of irreversibility, which is non-existent in a state of stable equilibrium. The arrow of time emerges from instability and nonlinearity; it is a source of phenomena such as self-organization and evolution. From this perspective, changes within a sociotechnical system do not appear at the end of a given period of time, but as the arrow of time appears, as it emerges from nonlinear processes within a non-integrable dynamic system. In other words, the notion of the arrow of time becomes 'meaningful' in irreversible processes.

[2] Please see Prigogine, I. (1984). Order out of chaos: Man's new dialogue with nature and Prigogine, I. (1980). From being to becoming: Time and complexity in the physical sciences.

Consequently, the very evolution of a system is its ability to draw out its arrow of time, passing through a state of unstable equilibrium and therefore, nonlinearity. This arrow of time is a driving force and constitutes a source of change within the very structure of the system. Thus, it only exists in existence, in other words, only following the appearance of irreversible processes leading to change and evolution, unlike a classical approach where 'time' appears before any ongoing, never-ending existence.

The evolution of a sociotechnical system does not, therefore, rely upon an ongoing, never-ending 'time' factor, as could be defined when using a modelling tool based on the Bertalanffy's general system theory. A sociotechnical system generates its own arrow of time, characterized by nonlinear and irreversible phenomena that highlight a break in the symmetry. Time only takes on a 'meaning' because a sociotechnical system is a complex system. Thus, the stability (even the safety) of a sociotechnical system can be obtained by modifications to its structure in a given context, in other words, by irreversible processes in response to a given context.

Systemic accident models consider time as a parameter that has no influence on the studied system. However, we have just seen that it is during the emergence of this arrow of time, in a state that is far removed from the equilibrium, that a system can build its own safety-oriented structure. But that is another story, for a completely different book!

Index

A
Accident models, 223–227
Accuracy, 113, 115
Alarm system, 75
  and system shutdown, 79
Analytical thinking, 3, 4
Approximate reasoning, 111
Asymptotic stability, 224
Auxiliary variable control, 78

B
Barendrecht project, 211
  decision making support, 219
  External Experts, 217
  hatched squares, 217
  methodology application, 212
  organizational structure, 213
  principal inadequate control actions, 215, 216
  proposed safety control structure, 216, 218
  risk management methodology, 219
  risks, interconnections, 215–217
  safety control structure, 213–215
  systemic risk management framework, 218
Bulk chemicals, 95

C
Calcination, 157
  in Novosol® process, 164–165
Capture, Transport and Storage of CO_2 (CTSC) technology
  Barendrecht case study (*see* (Barendrecht project))
  causal graphs, 208
  climate change mitigation, 197
  CO_2 leakage, 197
  control actions, controller, 207
  endogenous and exogenous controllers, 207
  financial/economic risks, 201
  global energy policy, 197
  hazardous state, 207–208
  health, safety and environmental risks, 200
  industrial CO_2 emissions, 197
  organizational and human risks, 201
  organizational structure, 201
  phases, 219
  policy/strategy risks, 199, 200
  public acceptance (social risks), 199
  regulatory/legal risks, 200–201
  risk assessment and management approaches, 198, 199, 201, 202
  risks and consequences, 204–205
  safety control structure development, 207, 208
  safety structure, 205–206
  sociotechnical system, 198, 201
  subsystem and project phase, 203
  system dynamics models, 208
  system safety constraints, 206
  technology scale-up problems, 208–211

F. Guarnieri, E. Garbolino (eds.), *Safety Dynamics*, Advanced Sciences and Technologies for Security Applications,
https://doi.org/10.1007/978-3-319-96259-7

CAST, *see* Causal Analysis based on STAMP
Causal analysis based on STAMP (CAST), 123
  accident analysis, 138–146
    dynamic phase, 144–146
    static and dynamic models, 138
    static phase, 139–144
  accident model, 123, 180, 181, 185, 190, 193–195
Causal graph, 97–99
  causal relations, 109, 118–119
  development, 96
  disaggregated, 44–45
  discharge and transfer system
    at aggregate level, 43–44
    disaggregated, 44–45
  generic model, of safety behaviour, 111–117
    accuracy, 113, 115
    application of, 118–119
    intelligent actions, 115
    motivation, 111–112
    self-efficacy, 113
    simulation, 115–117
    stress, 112–114
Chlorine
  accidentology, 35
  cells breakdown, 34
  discharge and transfer activities, 36–37
  on human health, 34, 35
  initial lethal effects, 35
  irreversible effects, 35
  physico-chemical, 34
  significant lethal effects, 35
  toxicological properties, 34
Chlorine discharge system
  auxiliary variables, 47, 48
  control unit, 48
  control variable, 48
  definition and function of variables, 49–50
  failure and systemic safety measures, dynamic analysis, 60
    advantages, 89–90
    disadvantages, 90
    French regulations, 62
    methods and tools, 88–90
    system failures analysis, 62–64
    using Stella, 64–88
  feedback loops, 46
  flow variables, 45, 46, 48
  model of, 49
  production flow, 48
  to production unit, 61
  stock variables, 45, 47, 48
Chlorine leak protection measure, 76
Chlorine line modelling, 50–52, 54
  flow and hot water requirements, 54
  liquid and gas fractions, 53
  risk analysis, 72–75
  variables used, 53
Chlorine sensor, 80, 81
Classified Installation for Environmental Protection, 32
Closed loop system, 127
Cognitive model, 9
Competence, 97, 115
Comprehensive failure analysis, 31, 32
Computer-based control systems, 8
Computer simulations, 12
  programming, 22
Contaminated sediments, *see* Sediment contamination
Control loops
  closed, 127
  open, 126
  STAMP model, 133–137
Control mechanism, 8
Control theory, 8, 13, 124–125
Current situation variable, 116
Cybernetic system, 8

D

Decision-making processors, 8
Decision model, 9
Disaggregated causal graph, 44–45
Discharge system and chlorine pipeline, risk analysis, 72–75
Dynamic analysis, safety behaviour
  generic model of
    causal diagram representation, 111–117
    fuzzy representation, 117–119
    fuzzy set theory, 110–111
    principles, 109–110
  industrial safety, human and organizational factor value, 107–108
Dynamic model design, 31, 144–146
Dynamic phase, CAST, 144–146
Dynamic risk analysis approach, 11
Dynamic system, 8

E

Emissions trading system (ETS) program, 212
Exponential growth oscillations, 224

F

Fatigue loop, 113, 115
Feedback loop, *see* Closed loop system
Feedforward loop, *see* Open loop system
First order system, 126
Floating production storage and offloading (FPSO), 179, 181, 183–184, 187–195
  offloading operation, 183
  in offshore oil and gas industry, 179
  safety control and management, 180
  technical intervention, 184
Forklift trucks stack products, 95
Forrester graphical language, 37, 38
Forrester's method, 13, 24
Fuzzy logic, 107
Fuzzy set theory, 108, 110–111, 117–118

G

General model of normal operations, 56–60
  chlorine transition flows, 58
  discharge simulation of chlorine wagons, 56, 57, 59
  liquid chlorine flow, 57
  state variables, 60
  variation types, 56
General system, 8
Generic model of safety behaviour, causal diagram representation, 111–117
  accuracy, 113, 115
  intelligent actions, 115
  motivation, 111–112
  self-efficacy, 113
  simulation, 115–117
  stress, 112–114
Genetics, emergence, 4
Genetic system, 8
Graph/stock-flow diagram, 12

H

Hazard analysis, 146–149
Hazardous elements, 125
HAZOPS approach, technical procedure, 170–172
Heat production modelling, 52, 54–56
Hierarchical control structures, STAMP model, 131–133
Human controls, 134, 135
Human errors, 96, 108
Human–machine interface, 136

I

Important safety elements (ISE), 81, 82
Inadequate control loop, 144
Incident tracking system, 27
Industrial Development and *Design–Purchasing*, 177
Industrial plant, 10
Industrial site and activity, 33–34
Intelligent actions, 115
Intelligent system, 8
Internal Operations Plan (IOP), 33
International Energy Agency (IEA), 197
International Maritime Organization (IMO), 185

L

Leakage sensors, 8
Linear causality, 225
Linguistic variable, 110

M

Meadows Report, 13
Method organized for a systematic analysis of risk (MOSAR) procedure, 5
Micro-organizational factors, 109
Modelling analysis
  industrial safety, human and organizational factor value, 107–108
  methods, 12
  safety behaviour, generic model of
    causal diagram representation, 111–117
    fuzzy representation, 117–119
    fuzzy set theory, 110–111
    principles, 109–110
  and simulation method, 12–22
  tools, 12
Model taxonomy, 9–10
Monomers, 95
Motivation, 97, 111–112

N

Non-bulk chemicals, 95
Normative model, 9–10
Novosol®
  contaminated residues, 162
  mineral residues, 161
  Novosol® flowchart, 162, 163
  physicochemical process
    calcination, 164–165
    dynamic model of phosphating, 165–168
    phosphating phase, 162–164

O

Open loop system, 126
Operator behaviour, 106
Organisation for Economic Co-operation and Development (OECD), 202

P

Peroxides, 95
PHAST® software package, 31
Phosphating phase
  dynamic model of
    in Novosol® process, 165, 166
    structural variables, 165, 167
    structure description, 165, 168
  gas cleaning system, 164
  in Novosol® process, 162, 163
Physicochemical process, Novosol®
  calcination, 164
  dynamic model of phosphating, 165–168
  phosphating phase, 162–164
Predictive model, 10
Preliminary risk analysis, 140–141
Problem articulation, 96–97
Process accidentology, 133
Process models, STAMP model, 133–137

R

Remediation processes of contaminated sediments
Risk management
  and categorization, 143
  for CTSC projects (*see* (Capture, Transport and Storage of CO_2 (CTSC) technology))
  and environmental impact, 25
  at industrial installation
    comprehensive failure analysis, 31, 32
    dynamic analysis, 32
    dynamic model design, 31
    failure consequences, comprehensive simulation, 31, 32
    prevention testing and protection methods, 32
    system behaviour simulation, 31
Risk prevention challenges
  chemicals, hazardous nature, 32
  hazard study, 33

S

Safety analysis process, 135
  STAMP model, theoretical foundations of, 129–130
Safety behaviour
  generic model of
    causal diagram representation, 111–117
    fuzzy representation, 117–119
    fuzzy set theory, 110–111
    principles, 109–110
  industrial safety, human and organizational factor value, 107–108
Safety control structure, establishment of, 141–142
Safety information system, 28
Safety management systems (SMS), 25, 26, 34
Second order system, 126
Sediment contamination
  management approach
    assessment of, 154–155
    calcination, 157
    environmental benefits, 156
    Environmental Protection Agency, 153
    human health and environment, 152
    management, 158
    methods, 158
    Novosol® process, 156, 158
    prevention and control, 155–156
    remediation process, 157
    sources detection, 153–154
    treatment of, 156–158
  treatment technologies, 158–160
Self-efficacy, 113
Simulation model construction, 11
Simulation modelling, 96
Socio-cultural system, 8
Socio-technical control structure, 131–133
Socio-technical system, 146, 147, 227
Specific response plan (SRP), 33
Stable equilibrium, 225, 226
Stable oscillations, 224
Stamp-based analysis (STPA)
  application, 151
  control structure and process models, 175–176
  hazard analysis, 167, 169–178
  risk categorisation
    control loop, maintenance and upgrade, 177, 178
    safety constraints, 176
    system's life cycle, 176
  safety control structure, 172, 174
  system requirements and constraints integration, 172–173
  system risks, preliminary analysis, 167, 170–172
STAMP model, *see* System theoretic accident model and processes (STAMP) model

Static phase, CAST, 139–144
STELLA software, 31
chlorine gas leakage modelling, 69–71
design and representation, 39, 40
to dynamic system model, 37, 64–88
Forrester graphical language, 38, 39
liquid chlorine leakage during discharge, 64–69
model of, 65, 66
simulation effect, 65, 66, 68
variables used, 65
modelled system, 40
preventive and protection measure modelling, 75–82
regulatory feedback loops, 40
risk analysis, discharge system and chlorine pipeline, 72–75
simulation and modelling tool, 22, 26, 41, 42
stock–flow diagram, 40, 41
stock variable interface, 41, 42
Stock–flow diagram, 96, 99–101, 115–116
STPA, *see* System-theoretic process analysis
Stress, 97, 112–114
Structural stability, 224
Subsea production system, 179
System behaviour simulation, 31
System dynamics, 12–22
application, industrial accidents, 26–29
application to supply chains, 24
automotive industry, 24
causality, 20–21
economic activities, 23
feedback loops, 15–17
industrial production structures, 23
in industrial safety, 13, 23
industrial system, 94–96
logistical unit, 24
modelling and dynamic analysis, 28, 96
causal diagram, 97–99
problem articulation, 96–97
simulations, 101–105
stock–flow diagram, 99–101
non-linearity, 17–20
principle characteristics, 22
reporting mechanism, 28
state and force, 14
stock-flow diagram, 28
structural change, 21–22
Vensim® and, 93–94
Systemic accident models, 223–227
Systemic approach, 2
vs. analytical approach, 5
complexity, 7, 107–108
comprehensiveness, 6
functional pole, 2–3
genetic pole, 3
industrial hazards, prevention and protection, 5
interaction, 6
interdisciplinary research development, 5
modelling approach, 7–29, 33
ontological pole, 2
organization, 6–7
Systemic risk management approach, CTSC projects, *see* Capture, Transport and Storage of CO_2 (CTSC) technology
Systemic safety management systems (SSMS), 25
Systemic safety model, 223
System requirements
definition of, 140–141
integration of, 142
System simulation with prevention and protection measures
atmospheric chlorine sensors, 85
chlorine flow, 84, 85
chlorine wagon discharge shutdown, 83, 84
small leak, collector, 86–88
variables representation, 82, 83
Systems theory, 13
System theoretic accident model and processes (STAMP) model, 13, 14, 123, 151, 179, 180, 198, 224
accident investigation internal report, 187
applied methodology, 182
business management, 194
CAST, accident analysis
dynamic phase, 144–146
static and dynamic models, 138
static phase, 139–144
component analysis, hierarchical structure, 190, 191
constraints, 181
controlling systems, 181
coordination and communication, 192
during loading operation, 192
downgraded mode of operation, 188
flow rate, 189
FPSO/crude oil transfer system, 181, 183, 184
government and environment, 195

System theoretic accident model and processes (STAMP) model (*cont.*)
  hierarchical control structures, 131–133
  hierarchical management of maintenance, coordination and communication, 191
  hierarchical safety control structure, 185–187
  management and production, 194–195
  process, 139
  Process Logical Controller, 183
  process models and control loops, 133–137
  safety constraints, 130
    in system and required configurations, 184–185
  Shut-Down Valve, 183
  STPA, safety assessment, 146–149
  system dynamics approach, 192–193
  tanker's Loading Master, 187
  technical infrastructure, 193
  technical installation failure analysis
    fibre optic cable, control and feedback, 189
    rupture disk breaking, control and feedback, 190
    SDV open, control and feedback, 190
  theoretical foundations of
    control and safety, 129–130
    control conditions, 127–128
    control loops, 126–127
    control theory, 124–125
System-theoretic process analysis (STPA), 123
  safety assessment, 146–149

T

Transcendental systems, 8
Triangulation, 2

U

United Nations Framework Convention on Climate Change (UNFCCC), 197
Unlimited growth, 224
Unpredictability, 7

V

Vensim®, 93–94, 99

W

Warning system, 77